D0569734

The BIOSPHERE

Also by Carl Heintze

The Bottom of the Sea and Beyond
Circle of Fire: The Great Chain of Volcanoes and Earth Faults
Genetic Engineering: Man and Nature in Transition
A Million Locks and Keys
The Priceless Pump: The Human Heart
Search Among the Stars
Summit Lake: Four Seasons in the Sierras

The BIOSPHERE
earth, air fire, and water

by **CARL HEINTZE**

Drawings by
Wayne Harmon

Library · St. Joseph's College
222 Clinton Avenue
Brooklyn, N. Y. 11205

THOMAS NELSON INC., PUBLISHERS
Nashville • New York

Copyright © 1977 by Carl Heintze

All rights reserved under International and Pan-American Conventions.
Published in Nashville, Tennessee, by Thomas Nelson Inc., Publishers, and simultaneously in Don Mills, Ontario, by Thomas Nelson & Sons (Canada) Limited. Manufactured in the United States of America.

First edition

Library of Congress Cataloging in Publication Data
Heintze, Carl.
The biosphere.
Bibliography: p.
Includes index.
1. Earth. 2. Atmosphere. 3. Hydrology. 4. Pollution.
I. Title.
QB631.H44 551.4′6 76-51231
ISBN 0-8407-6498-7

for Marge
with Love

92638

CONTENTS

The BIOSPHERE

The BIOSPHERE

Stand on a beach and look at the sea. The sky stretches in airy splendor overhead until it meets the horizon. The sea is a restless march of endless waves advancing to crash on the shore. Both sea and sky seem immense and mysterious. Beyond the sky lie the moon, the stars, and space. The ocean is deep and dark, unyielding to most attempts to see its bottom.

To the ancient Greeks, both sea and sky were parts of the four elements they believed made up the universe: earth, air, fire, and water. Today we know there are more than one hundred elements in the universe, yet it is still possible to express our world with the four divisions of the Greeks. We see the world's land surface as earth. Water is the ocean, rivers, and lakes. Fire is the sun and oxidation, the burning of combustible materials. Air is the atmosphere we breathe.

Taken together, these four elemental states make up the biosphere, the part of the world in which we live. They include the thin film of gases and liquid that coats the surface of our planet. All life in the biosphere exists on the surface of the earth, in its air, or in its waters.

Yet the biosphere also is remarkably limited. Only a few

miles in depth, it spreads over the earth's surface in great disproportion to the rest of the planet. Its depth or size, however, is not as important as life's ability to move laterally at great distances through it.

Further, the biosphere is both a shield and a bank. It allows most sunlight to reach plants and animals, yet it also shields them from harmful radiation from space. The biosphere also nourishes life by storing within it the sun's energy.

Were it not for these qualities, there would be no living creatures, no weather, no waves, no sea, nor any of the beauty found on earth.

Earth, air, fire, and water all are necessary to the biosphere. Each is necessary for the action of the others. Without one, the others might well not exist. Certainly without all of them, we could not survive. Instead the earth would be as lifeless as the surface of the moon or the arid, dust-blown plains of Mars.

How did the earth come to have a biosphere when the other planets of the solar system appear to have none? How did life come to inhabit it? How does the biosphere work, how long will it continue to shelter life, what is its future?

This book deals with these questions, but does not attempt to answer them all. Indeed, not all the answers are known. Perhaps someday we may be able to answer some of the unresolved mysteries about the living world.

—1—

EARTH

The biosphere was created long ago, so long ago that little of its birth can be described with any accuracy. The events that brought it into being can only be guesses, estimates derived from very recent observations in man's past history.

The precise age of the biosphere is unknown, but it must date to a time after the birth of the sun. The sun is an ordinary star, a second-class star among the many stars in the heavens. It is a part of the Milky Way Galaxy, a great aggregation of billions of stars moving together through space like a giant pinwheel.

The sun is in one of the far reaches of one of the arms of the galaxy, traveling in an orbit which will take 200 million years to complete. With it travel the earth, the moon, the other eight planets of the solar system, their various satellites, and bits and pieces of broken material called asteroids and meteorites.

Except for the nine planets of the solar system, we have no knowledge of any other planets in the universe. None are likely to be discovered in your lifetime. Even if they were, the distances between stars are so great it seems unlikely we will know whether or not they carry life on their surfaces.

It also now seems almost certain no other life will be found on any of the sun's planets, that the earth and its biosphere are unique among the worlds which revolve around our star and that the creatures of earth may be the only intelligent beings mankind will ever know.

All this does not mean that either life or biosphere is unique in the universe. It would be presumptuous to believe that the earth is the only planet capable of sustaining life. It is far more likely that other planets with life on them do exist somewhere in space, and one of the great challenges before mankind will be the search to find them. Until that happens, we must search for the beginnings of the biosphere on our own earth and the other planets of the solar system.

It is difficult to generalize about the planets. Although they have many similarities, they also have important differences. In general, they can be divided into two groups: the inner or minor planets—so called because of their size—and the major or outer planets.

The earth, Mercury, Mars, and Venus are considered the minor planets. Jupiter, Saturn, Neptune, Uranus, and Pluto are the major or outer planets. As their name implies, the inner planets are much closer to the sun than the outer planets. All also are in orbits about equally distant from one another. All four of the inner planets are quite small compared to the major planets, many times smaller. Even so, the minor planets are not uniform in size. Mercury and Mars are smaller than the earth and Venus.

The earth alone, however, has a satellite, the moon, which is proportionally larger to its mother planet than any of the others, so large that the earth and moon are almost a double planet. Neither Mercury nor Venus have moons. Mars has two,

M81, a constellation in Ursa Major, similar to the Milky Way Galaxy in which the solar system exists. The earth is revolving around the sun in one of the far arms of the Milky Way Galaxy.

Courtesy Lick Observatory, University of California at Santa Cruz

but both are quite small, much smaller than our satellite, so small that for a time they were not detected by astronomers.

The most obvious difference between the inner and outer planets is their size. All the outer planets are much larger than the inner planets. One, Jupiter, is immense—a huge ball surrounded by twelve moons. Another, Saturn, seems surrounded by rings—thin disks of ice or dust. Except for Pluto, the outermost planet, all the major planets have satellites. Pluto may once have been a moon, which escaped from a major planet to travel a lonely and erratic orbit at the edge of the solar system.

Except for Pluto, all the outer planets have atmospheres, but each is different from that of earth. If Pluto has one, it has not been detected. The atmospheres of the outer planets seem to be made of thick, very cold, perhaps frozen gases. Mars, the earth, and Venus all have atmospheres, Mercury has none. But there are important differences between the inner planets' atmospheres.

The atmosphere of Venus is rich in carbon dioxide, but lacking in free oxygen. The earth's atmosphere is composed mostly of nitrogen, free oxygen, and carbon dioxide plus a great deal of water vapor as well. Mars has an extremely thin atmosphere in which there is apparently little or no free oxygen and almost no water vapor. The Martian atmosphere, however, does contain carbon dioxide. Mercury has no atmosphere at all and, like the earth's moon, lies exposed directly to space and the fierce rays of the nearby sun. Excepting earth, none probably hold life.

Yet despite these differences, the minor planets and the major planets, too, have some important characteristics in common. All are round or nearly so. All revolve around the sun in orbits that are slight ellipses—that is, paths which are not perfectly circular. All travel around the sun in approximately the same plane, one which is roughly parallel with the sun's equator. All also revolve on their own axes, but each has a different period or length of rotation. Jupiter, for example, turns around almost once every eleven hours. The earth revolves once every twenty-four hours. Venus seems to be turning in a direction opposite to the rest of the planets, but very slowly, so slowly that it takes many earth days to complete a single rotation.

Why are the planets turning? Why are they moving around the sun?

This is an elementary question about the solar system. Indeed, it is an elementary question about all bodies in the universe. Stars and planets all are moving in relation to one another, but their movement has puzzled men over many years.

All agree some force is driving planets and stars through space, but its nature is still a mystery.

We call this force gravity, but we do not really know what it is. Gravity was first appreciated and reasoned into laws two centuries ago by Sir Isaac Newton, the famous English philosopher-scientist. It is doubtful that Newton devised his laws of gravity after being hit on the head by a falling apple—as legend has it—but he did perceive that all bodies in space are attracted toward one another. From this he framed the concept and the "laws" of gravity.

Newton concluded any object would move through empty space with a speed equal to the force originally used to propel it. He also deduced that if two bodies collided in space, they would fly apart with a force equal to that which drove them together, and he worked out equations which seemed to fit all these situations. Later studies have shown the formulas work well with large bodies moving through space, but cannot be applied to atomic particles. Nevertheless, they have proved helpful in trying to understand the movement of the planets.

About 5 billion years ago, most modern scientists believe, there was no sun or planets, only a mass of gas swirling about in space. This great cloud of gas probably was composed of hydrogen atoms. Hydrogen is the lightest and most abundant element in the universe. It also is the simplest, consisting of a single electron revolving around a single proton nucleus.

For billions of years, this gas cloud moved around in space without taking definite form. Then it condensed. Its condensation probably happened because of gravitational attraction between atoms in the plasma. Whatever the cause, the hydrogen atoms in the cloud were packed tightly together, so tightly their constant collisions began to heat the center of the cloud to very high temperatures, millions of degrees hot. At the same time, the pressure inside the cloud became immense.

As billions of hydrogen atoms collided with one another in the intense heat and pressure, their electrons were stripped away, and two hydrogen atoms were fused together to make a single helium atom. A small amount of energy was left over from this fusion reaction. Multiplied by billions of such events, it lighted up the cloud, and the sun was born.

Not all the gas in the cloud was compressed into the sun, however. Some outer eddies and currents in the cloud swirled together to form the beginnings of the planets. The planets formed in space at some distance from the sun and were held in orbits because gravity balanced one mass against the other. The planets were not large masses of material as compared to the sun, but within them heavy elements were formed and pulled together, again by gravity, to make their cores. The smaller size of the inner planets is an indication that they were able to balance their orbits only by getting rid of most of their light gases.

The outer planets retained theirs. All else in the early solar system was swept away into space. It is probable, however, that some bits of heavier material did not immediately become part of the forming planets. Instead, these were picked up as the planets orbited the sun, somewhat like a small snowball increasing in size as it is rolled across a field of snow.

The impact of these smaller pieces as they fell to the planets' surfaces created the great craters still found on the moon, Mercury, and Mars. The earth may once have had such craters also, although, over billions of years, our planet's surface has been so eroded by wind and water that they have disappeared.

Each new planet had an atmosphere, with the exception of Mercury, which must have lost its gases to the sun. But for reasons not clear, the atmospheres of the inner planets came to differ substantially from one another.

Library · St. Joseph's College
222 Clinton Avenue
Brooklyn, N. Y. 11205

The earth also developed a magnetic field. Some other planets have magnetic fields around them—Jupiter's, for example, is more intense that that of our world—but some may not.

The earth's development took place at an ideal location, just the right distance from the sun so that water could be retained on its surface. Had the earth been closer to the sun, it would have lost most of this precious substance. Some scientists reason that were the earth about 6 million miles closer to the sun, it would be a far less hospitable place for life. The earth also was exposed to the proper amount of radiation from its parent star, enough to make it warm but not too warm.

Both this heat and water have proved essential to the development of life on our planet.

Without the proper amount of heat from the sun or without water, life could not have come into being. It is these two requirements that make it unlikely that life either arose or survived on the outer planets of the solar system. On Venus, so much carbon dioxide remained in the atmosphere that water could not form, and it remained a cloud-shrouded hothouse too hot for life either to begin or to survive. On Mars, free oxygen may once have been more abundant than it is today, but Mars may, in the end, have been too far from the sun to receive sufficient light for living plants and animals. Yet it remains the only other planet likely to have life on its surface. Pictures of it taken by Mariner space probes show what seem to be riverbeds, now dry, which may have been once sculptured by the action of moving water. Mars remains an unlikely home for life, however. The probes landed on its surface in 1976 by the United States have turned up no definite proof that it contains even microscopic forms of life.

Studies of the Martian atmosphere from earth show that its carbon dioxide collects as a layer of frost on the planet's north and south poles, but no unmistakable signs of water have been discovered. Only on earth was water formed. Exactly how water came to the earth is as mysterious as the question of what happened to any that may have been on Mars. The earth probably has not always been as covered with water as it is today, but early in its history the oceans began to appear, spreading across its surface until they occupied three quarters of it. Only a world island remained. This island, later to be broken apart by geologic forces from within the earth, was at first barren of life. Not a tree, a blade of grass, an animal, an insect, or a bird existed in the biosphere.

Instead, life—the vital essence which makes the biosphere of our planet different from all others in the solar system—evolved in the seas. As with the beginning of the biosphere itself, the details of life's creation are uncertain. But the first life must have been created from the basic, abundant elements on the earth's surface: oxygen, carbon, hydrogen, nitrogen, phosphorus and other "light" substances.

These elements were and still are in the seas or a part of their waters. Some form of radiant energy also had to be added to create life. It may have been the rays of the sun, it may have been cosmic rays from beyond the solar system, or perhaps the continued lightning which bombarded the seas from the clouds above them. However, energy was transmitted to the oceans, and in them dioxyribonucleic acid (DNA) was formed—the molecule capable of reproducing itself and the most important ingredient of all living organisms.

Over millions of years the organisms created by reproduction gradually grew more and more complex in structure. At the same time, the earth's atmosphere and oceans began to change. They became cooler and more stable. The life within the seas seized this opportunity to move to land. There were advantages to living on land not present in the oceans—easier movement, for example. Aquatic plants became the first to realize this possibility. Migrating from shallows along the coasts of the continents, they gradually moved inland to cover the land surface of the earth. Plants could live and grow by drawing carbon dioxide from the air and energy from the sun. The by-product of this process was the release of oxygen into the atmosphere. As more and more plants reached shore, more and more oxygen was released into the air, gradually making it possible for animals, too, to leave the seas and come to dwell on the continents. The plants already there provided them with energy, and the additional amounts of free oxygen in the air allowed them to breathe.

The first land animals were amphibians, creatures who spent part of their lives in the water and part of it on land. Amphibians must have evolved from fishes and fishes from even more primitive creatures, perhaps those with stony shells or with bodies like the jellyfish of today.

Finally, amphibians abandoned the seas altogether and became the first completely land-dwelling creatures in the earth's history. They became a part—a key part—of the carbon dioxide–oxygen cycle of today by which plants and animals nourish one another. At the same time, plants remain the key to animal survival on earth. Without their ability to capture the energy of the sun, an ability not shared by animals, there would be no animal life on our planet, perhaps no life at all.

Thus, the biosphere is a fragile world, dependent on a number of delicately balanced factors: on the interaction of plants and animals; on the constant, steady, radiant heat of the sun; on the widespread availability of water; on the distance of the earth from the sun; and on the continuing evolution of living systems. For life does not exist in spite of the biosphere, but because of it. Life is dependent on the biosphere, just as the biosphere is dependent on life for its future. Our planet's surface is a complex, highly integrated place, in which each living system has a part. Though it probably is not unique in the universe, it seems without a duplicate in the solar system.

—2—
AIR

The air is the part of the biosphere with which we are most familiar. Almost every moment of our lives is spent within it. We breathe its gases, walk, fly, or drive through them. Our world is seen from its depths. Yet most of the time, we forget we live at the bottom of an ocean of air. So familiar and so all-enveloping is it, we simply ignore it. Part of our forgetfulness may be because the air is odorless, tasteless, and impossible to feel. If we "smell" the air, it is not the air that we are sniffing, but its contaminants. If we "feel" the air, it is its heat or water vapor that affects us, not the air itself. If we "taste" the air, we are tasting airborne substances.

Yet the air exists. If it did not, the delicate system by which we breathe and through which we are able to use the gases of the atmosphere would fail. Nor would there be a single blade of grass, tree, or any other form of life on the earth. The existence of the atmosphere and the constant recycling of its gases makes it possible for living organisms to survive in the biosphere.

Although it is a gaseous ocean, with currents, eddies, and other similarities to the seas which cover the globe, the atmosphere obviously is different from the earth's water cover. The air is not as uniform as the oceans. Its concentration and weight varies with altitude from sea level and with the seasons of the year.

Variations in the atmosphere take place because the air is a combination of gases, rather than a liquid. The concentration of any gas is dependent on the closeness with which its atoms are packed together. The atoms of the gases in the atmosphere are held together by gravity, the force which affects all the universe. Gravity's pull is greater on the atmosphere close to the surface of the earth, hence the concentration and pressure of gases is greatest closest to the earth's surface. As gravity diminishes with altitude, the pressure of the atmosphere grows less. At sea level the atmosphere exerts 14.7 pounds of pressure for each square inch of surface, but this rapidly diminishes with altitude, so that at even a mile above sea level, some persons have difficulty in breathing.

The temperature of the air also varies from point to point above the earth's surface. Air temperatures are dependent on a number of factors—altitude, the amount of water suspended as vapor in the atmosphere, and the speed with which air is being circulated. At sea level, the temperature of the air rarely falls below 60 degrees below zero or rises above 120 degrees Fahrenheit. Air temperatures tend to decrease through the lower levels of the atmosphere at about 1 degree Fahrenheit for each 100 feet of altitude.

These are only generalizations, however, and other factors are important in determining air temperatures at higher levels of the atmosphere. Insolation (see Chapter 3), also affects air temperatures, making the air generally warmer at the equator and colder at the poles, because the sun's rays strike the earth at greater angles at higher latitudes north and south of the earth's middle.

Water also is a key factor in the temperature of the air. The oceans absorb heat from the sun during the day and release it during the night. Land, on the other hand, reflects the sun's rays during the day, making differences in both day and night-time temperatures over land surfaces. In either case, the sun has an important effect on the temperature of the air. Water suspended in the air as vapor allows the air to absorb heat and also to release it as the water is turned into clouds, fog, snow, and rain.

Temperature is one of the ways in which scientists have been able to divide the atmosphere into levels. The lowest level of the atmosphere, the troposphere (from the Greek word *tropē,* meaning the act of turning), a region which extends from the earth's surface to 40,000 or 50,000 feet, is the home of all life, almost all clouds, the majority of the water vapor in the biosphere, and almost all the changes in conditions we call weather.

The upper limit of the troposphere is also about the limit to which jet aircraft can fly, and thus it is the practical limit to which most humans travel in the atmosphere.

Above the troposphere lies the tropopause, a dividing layer of air which separates the troposphere from the stratosphere. The stratosphere extends upward for from 100,000 to 150,000 feet above the earth's surface, depending on the time of year and the part of the earth over which it lies. Both the stratosphere and troposphere may vary in thickness, depending on whether or not the earth is enjoying summer or winter, being thicker at lower temperatures.

A few high clouds sometimes are found in the stratosphere, but usually it contains very little water vapor and low gas concentrations. The stratosphere also is the home of the Aurora Borealis (the northern lights), the gatherings of charged electrical particles. In the northern latitudes of the earth, these often appear as great glowing bars of light at night.

Vestiges of the troposphere's gases also are found in the stratosphere, but they are at such low concentrations that they could not support living creatures. In the higher layers of the stratosphere, the concentration of gases grows less, for there is less gravity to hold the atoms of gas together. The temperature of the gases in the stratosphere may grow higher, however. This also is true in the thermosphere and exosphere, the two highest levels of the atmosphere. There air temperatures may rise to more than 1,000 degrees Fahrenheit, but because gas atoms are so far apart there, little of this heat is transmitted from one atom to another. The thermosphere and exosphere do have sufficient atoms within them, however, to create friction on meteorites as they fall toward the earth from space, usually turning them to cinders before they reach the lower levels of the atmosphere. The upper levels of the atmosphere thus provide the earth's surface with enough protection to prevent

The levels of the atmosphere separated by height from the surface of the earth. Each level is differentiated from the one below it by a "pause." The levels vary in depth, depending on the time of year. Life exists only in the lowest level, the troposphere, which is also the home of most changes in the weather.

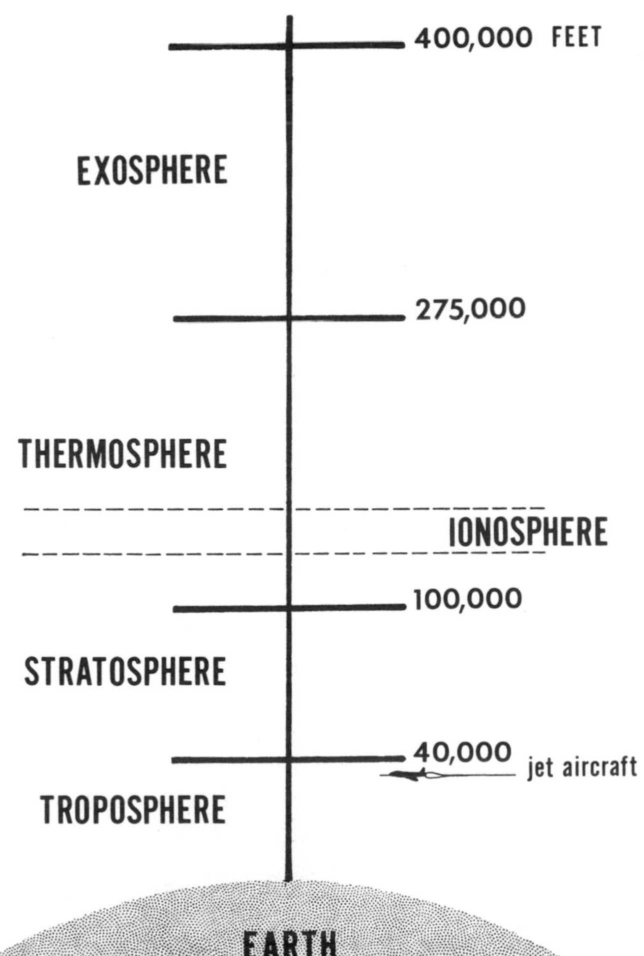

wholesale bombardment of the surface and its life by these travelers from space.

Water vapor in the atmosphere is "expressed" or made visible in some form of precipitation—clouds, fog, rain, sleet, hail, or snow. In any of these forms, it becomes visible when it is sufficiently condensed to allow water droplets to form in the air and fall toward the earth. The droplets condense around dust particles in the air. The additional weight of the dust particle also makes them fall toward earth more rapidly than if they had been formed only from water.

Whether it is visible, as clouds or other forms of precipitation, or invisible, as water vapor, water makes up a sizable part of the atmosphere, perhaps as much as 40 million tons. Much of it is invisible. Thus, simply because the air is clear is no guarantee that it is also dry. Clear air is as capable of carrying water as is cloudy or foggy air.

Water vapor usually is visible from the surface, however, as clouds. Clouds are important not only because they bring moisture to the land surface of the earth, but also because they make it possible to follow patterns of air movement. The movement of the air over the surface of the earth is complex, because air travels not only horizontally, but vertically as well, and because the gases of the air are constantly in motion. Air is never completely stagnant. The air which exists over one part of the earth today will be in a completely different place by tomorrow.

At lower levels of the atmosphere, it is often possible to chart air movement by watching how dust, smoke, and the products of volcanic eruptions are pushed this way and that by air currents.

Clouds, however, are the chief way in which men have been able to chart and predict the weather, as the changes in air movement and precipitation are called. The formations of clouds can tell a great deal about what to expect from them. Indeed, with enough information, even nonscientists can make rough but fairly accurate predictions about the weather from cloud formations alone.

In part this is because air currents in the atmosphere follow generally predictable pathways over the earth's surface. Any study of air currents must begin with the fact that the earth is a turning sphere. Not only does the air move around our planet, but the earth also turns beneath the air. Thus, even if air stood still, it would not be over the same point on the earth's surface for very long. The earth's rotation and the movement of the air help to create an apparent whirling of the air into circular pools called cyclones or anticyclones, depending on which direction they are moving. In the Northern Hemisphere, cyclones seem to be moving in one direction. In the Southern Hemisphere, they seem to be moving in the opposite direction. In either hemisphere, the apparent circular motion is called the Coriolis force, in memory of Gaspard Gustave de Coriolis, a French scientist of the nineteenth century.

An equally important factor in the movement of the air is the sun's heat.

Heat does not reach the atmosphere in uniform amounts, nor is it reflected or absorbed in uniform amounts by the ocean. In general, however, both the air and water around the earth's equator receive more heat than do the air and water near the poles. The equatorial air rises in an effort to cool and because it has more pressure within it. Because the air over the temperate zones of the world is cooler, equatorial air flows both north and south over the earth toward it. As the air grows cooler, it begins to sink back toward earth again, first at about 30 degrees north and south latitude from the equator, and then at about 60 degrees north and south latitude.

As the air reaches these two areas, some of it flows back toward the equator again, forming a constantly moving convection cell of air, but some of it remains aloft and continues toward the poles. That which finally does reach the poles grows cold and sweeps back along the earth's surface toward the temperate zones and the equator.

These six convection-cell movements make the general circulation of air over the face of the earth, but they are not as neatly defined as this description would indicate. Rather, they

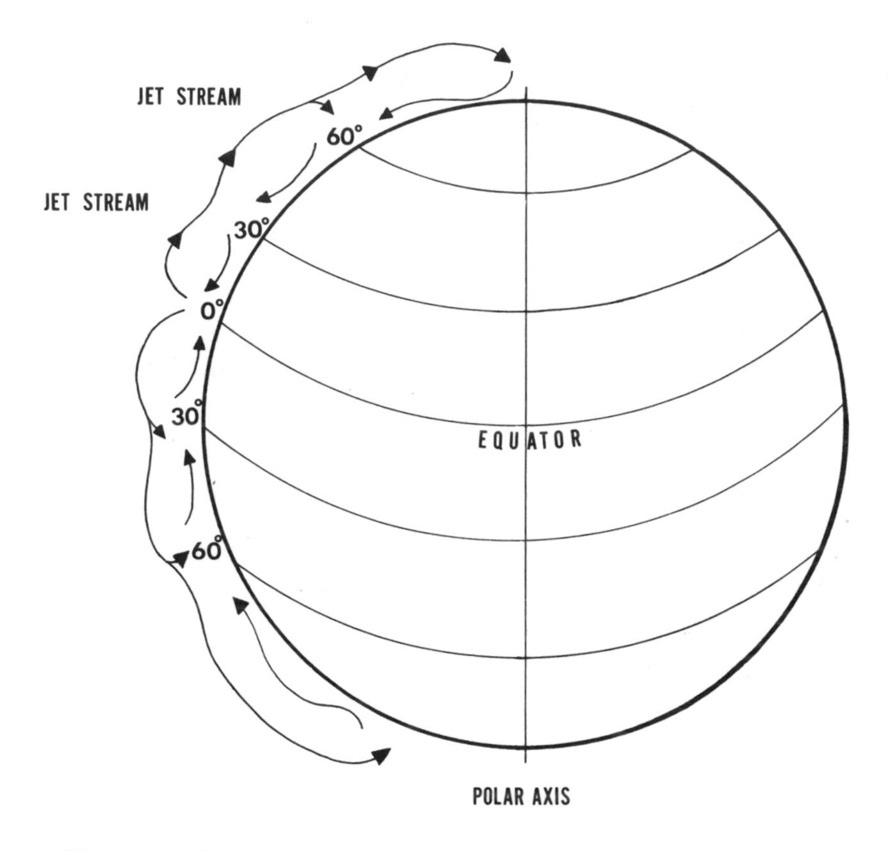

The general pattern of air circulation in the atmosphere, showing the convection cells. Only one side of the earth is shown. The cells cover the earth's surface and tend to vary in height according to the season of the year.

are general patterns of circulation, and their general movement helps greatly in explaining the regular direction in which winds on the earth's surface flow and how weather at higher elevations in the atmosphere is created. One always has to keep in mind, too, that the earth is turning beneath the cells as they move, adding an apparent direction to the wind.

The cells are the explanation for several wind phenomena: the trades, which blow in a constant direction across certain latitudes of the Pacific and the Atlantic; the doldrums, intervals between the cells at the earth's surface where winds often are slight or almost nonexistent; westerlies; and the strong surface winds at the poles.

More important, where the cooled higher-altitude air begins to sink earthward, the jet stream forms. The jet stream is an intersection between two cells where the vertically moving air of the cells creates a stream of very swift air at high altitudes in the troposphere. Usually the jet stream travels at several hundred miles an hour. The stream affects North America by blowing across the northern Pacific to British Columbia or Washington State. It then dips down over the United States. Coincidentally, jet aircraft traveling from west to east across the Pacific often "catch" the jet stream and use its tailwinds to reduce their fuel consumption. The stream is not named for them, however, but for the fact that it is a thin jet of fast-moving high-altitude air.

Jet streams seem to affect the location of cyclones and anticyclones as they travel across the earth's surface. Both tend to follow the path of the jet stream, and hence its location on the earth is an indication of where storms or fair weather are likely to occur. The jet stream is not always in the same place. During the winter months in the Northern Hemisphere, it tends to shift southward to bring winter storms to the United States, but sometimes this does not happen, and abnormal weather results.

Cyclones and anticyclones result from vertical movements of air along the jet stream's path, interacting with its horizontal movement. As warm air forms near the surface, it has greater pressure than cold air and begins to rise. It also expands. Such a body of air is called a "high" because its pressure is higher than the air surrounding it. If it is colder than the air around it, with less pressure and is sinking, it is called a "low."

In general, highs are an indication of fair weather—that is, weather without rain, fog, or snow—whereas lows may mean stormy weather is coming. The situation with regard to highs and lows is more complicated than this, however. It is not so much the appearance of a high or low that indicates stormy weather as it is the interaction between them. When highs and lows meet or come near enough to one another, air flows be-

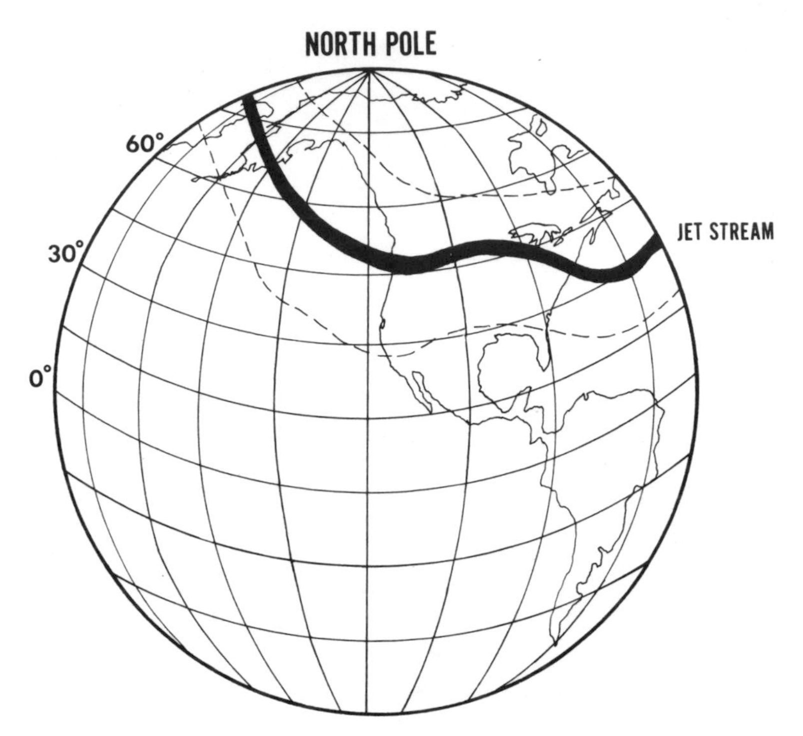

The pattern of the jet stream that crosses the Northern Hemisphere, with its boundaries. The stream varies upward or downward on the earth, depending on the season of the year and other factors, to the limits shown by the dotted lines.

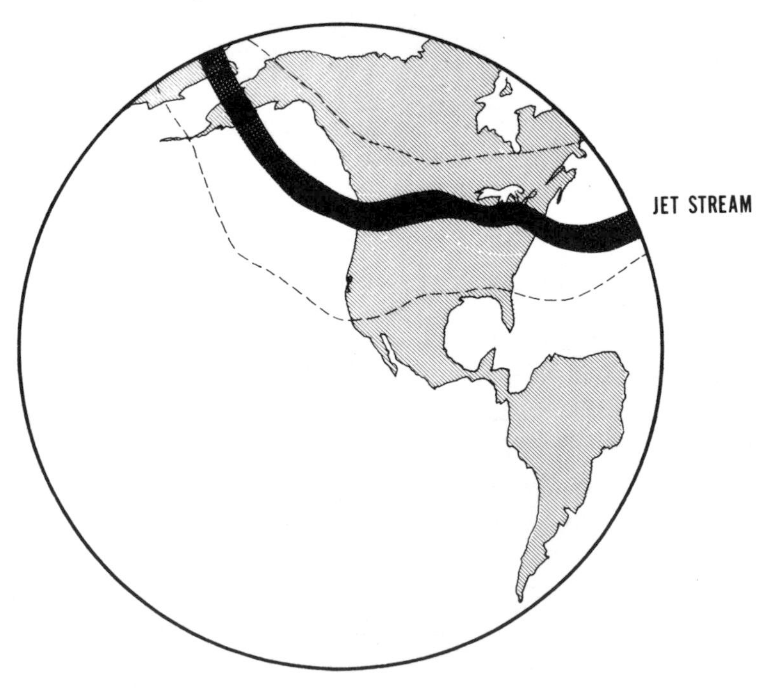

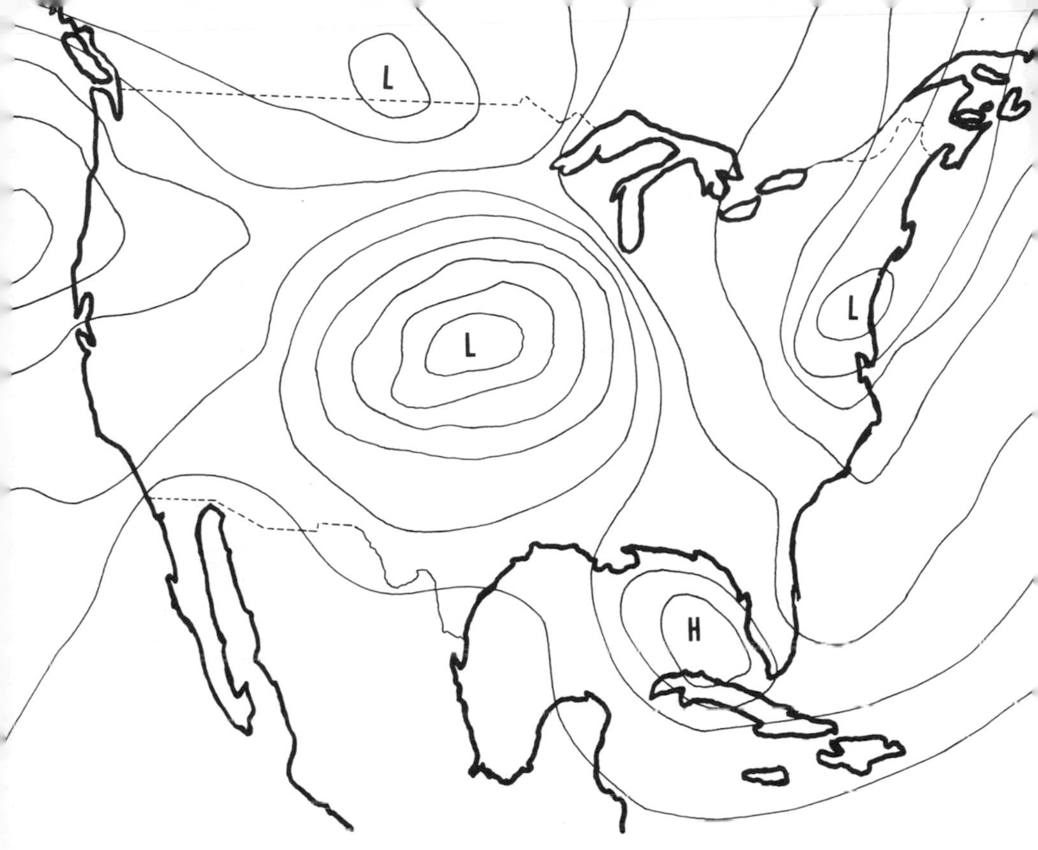

A pattern of highs and lows over the continent of North America, which can be taken as typical of the spring or summer of the year. The pattern is similar to many weather maps prepared from satellite photographs, but does not show wind velocities or barometric pressures.

tween them, and the water vapor in the highs tends to be condensed into precipitation.

Highs and lows interact with one another partly because of the stability or instability of air. Air is always moving in relation to the earth's surface, but it is not always rising or sinking. The rise or fall of pockets of air in the atmosphere takes place because of differences in the moisture content and temperature of such bodies in relation to the air around them.

Warm, moist air tends to rise. Cold, dry air tends to sink. The rate at which air rises or falls also is dependent on whether or not the air is cloudy or clear. Cloudy air tends to cool more slowly than clear air, because it contains more water vapor and hence may continue to rise for a longer time than clear air. This is particularly true of the clouds of some thunderstorms, for example, which may soar thousands of feet into the air very rapidly before they are sufficiently cooled to become stable air.

It is also important to remember that movement of the air always takes place in relation to the earth's surface and that the earth is constantly turning. Thus, even as the air is being churned around by differences in its temperature and water content, it also may be passing over a different place on the earth's surface. Highs and lows seldom remain stationary, but move

A thunderhead showing the immense tower of upward-thrusting air such clouds generate as they rise. Rain is falling from the base of the cloud and the air inside it is exceedingly turbulent and could be very dangerous to aircraft passing through it.

Courtesy National Oceanic and Atmospheric Agency

over the earth, constantly changing their position. There are, however, important exceptions to this rule. For example, during the summer in the Northern Hemisphere, a high generated by warm air rising from the sea lies almost stationary over the northern Pacific Ocean. Called the East Pacific High, it prevents storms moving down from the Gulf of Alaska from reaching California. Instead, they are shunted across Washington, Oregon, and British Columbia.

In the winter, the East Pacific High usually shifts southward, and rain and snow reach California. A similar semipermanent high can also be found over the Gulf of Mexico south of the United States. Its interaction with cold air from Canada moving down over the Plains States is responsible for much of the weather in the central United States.

The collision of warm and cold air in highs and lows makes weather in one of three ways. When warm air rides out over a cold low, it tends to hold it in place—because the cold air cannot rise. This sometimes creates a kind of lock on the cold air and makes an air inversion. But more often, it causes some of the warm air to condense into rain or snow. This formation is called a warm front.

The reverse also may be true. Cold air from a low may overwhelm a body of warm air, creating a cold front. Where the warm and cold air come together, rain or snow may form. Finally, cold and warm fronts may mix to create an occluded front. Precipitation also may form in places along an occluded front.

To find highs and lows and hence to discover the possibility of interaction between warm and cold fronts, weather scientists must have some idea of the pressure and temperature of air within specific places in the air. To measure air pressure, it is necessary to have a standard against which to check it, and this is the pressure of air at sea level, 14.7 pounds per square inch.

Most barometers, the devices used to measure air pressure, are calibrated at this figure. Some are made of glass tubes filled with mercury, whose columns are graduated from sea-level pressure, but some may be small tanks from which air has been

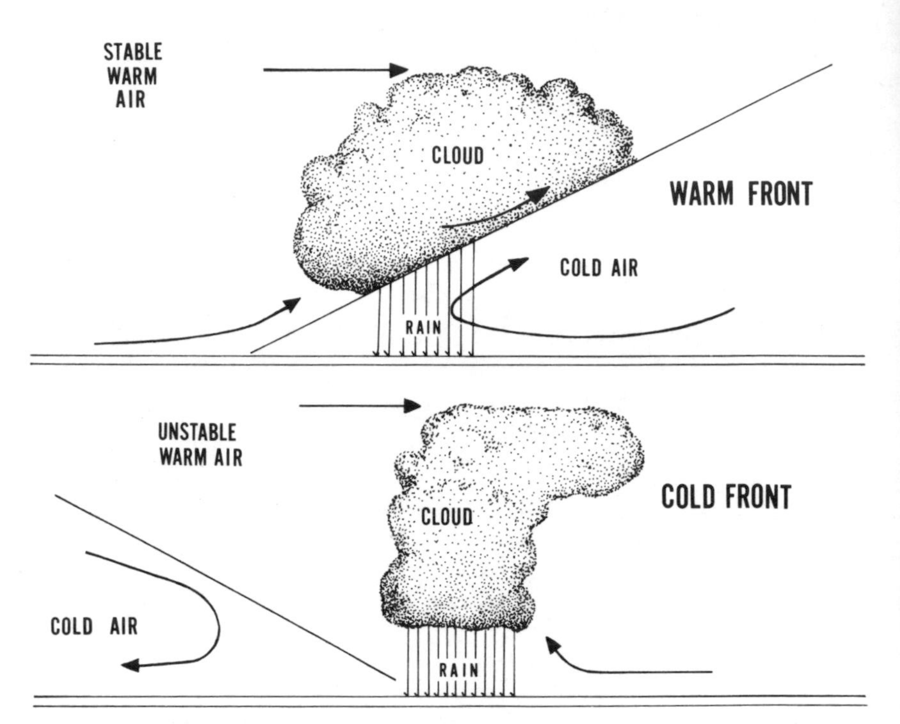

The general pattern of warm and cold fronts in the weather. (An occluded front is not shown.) These are idealized examples, and each warm or cold front in reality would look similar, but would be different in some respects.

removed, making it possible for the tank to reflect changes in the air pressing on it. In either case, barometers detect small but important changes in air pressure. Because they do, they make it possible to tell whether air pressure is rising or falling. In general, a falling barometer and a declining amount of air pressure indicate storms are likely to be on the way. A rising barometer means fair weather or clearing after a storm.

Barometer readings, of course, are only one of many bits of information weather scientists need to predict the weather. They also must measure the temperature of the air, the direction of the wind, its force, and many other factors that may affect weather changes. Technologically advanced nations of the

Hail descending to earth after having been precipitated aloft from a thunderhead. The mottled-dark-and-light clouds at right are typical of the kind generated by this kind of interaction between warm and cool air.

Courtesy National Oceanic and Atmospheric Agency

world over the past half century have developed thousands of weather-observation stations to gather such information. Less-developed nations have few such places and less ability to tell how the weather is going to change.

A vast amount of such information now goes into weather forecasting. It is gathered from ships at sea, balloons flown into the atmosphere, weather satellites, and ground measurements. Since the invention of computers, it has been possible to use such data to gain an ever-clearer picture of what is happening to the air, whether the air is visible as clouds or is invisible.

A weather-satellite photograph of the United States showing the cloud cover over much of North America on a typical summer day. The lines denoting state boundaries are superimposed on the picture as reference. Such photographs are of great help to meteorologists in predicting what the weather will be.

Courtesy National Oceanic and Atmospheric Agency

It also should be clear that even though masses of cold and warm air are constantly shifting around in the atmosphere as the earth moves beneath them, the shape of the earth's land surface also has important local effects on the weather. Ranges of mountains tend to force air upward. Mountain passes leak air into regions beyond the mountains. The oceans and inland bodies of water absorb heat. Deserts reflect it back into the air.

For example, the west slope of the Sierra Nevada Mountains of California receives a great deal of rain and snow during the winter, because moisture-laden winds from the Pacific blow in toward them and are pushed up by their rising shape, causing the winds to condense and drop their moisture before they move on over the deserts of Nevada and Arizona. The air which reaches beyond the mountains is literally wrung dry by this local effect.

Similarly, the eastern sides of the Hawaiian Islands are the precipitating force which causes the moisture of easterly winds over the Pacific to be dropped as rain on the islands' windward side. The island's land surface is sufficiently steep and high to do this.

Usually, there is also a pattern to the movement of air over coastal regions of the continents. During the day, the land radiates its heat into the air, causing it to rise and flow toward the ocean. During the night, the ocean, which has been absorbing solar heat in daylight hours, is free to release the heat into the now cool air above it, and the flow of air reverses itself and moves back toward the land. We know these air movements as on- and offshore breezes, and usually they are as regular as the night and the day.

Much the same kinds of forces are often at work between mountains and valleys. During the day, air warms more rapidly over the surfaces of the mountains than in the valley, but it also cools down more rapidly there during the night. As it cools at night, it flows downhill into the valleys below. This causes the warm air the valleys have been holding to be pushed aloft and carried toward the mountains. Thus there is set up a strong up-and-down flow of air between mountains and valleys, particularly where there are large differences in altitude.

In some parts of the world at particular times of the year, however, this flow may be constant in one direction for days at a time. When this occurs, it is called a foehn wind, a word taken from the German name for warm, constant winds that sometimes blow down from the Alps in Europe. But foehn winds also blow in other places. In southern California, they are called Santa Anas. In the Pacific Northwest, such winds are called chinooks. In Argentina, they are called the zonda, and in France, the aspre.

All such winds are regular air movements, ones which may take place every day, but there are some irregular winds which fit in none of these categories and for which there is no easy explanation. The most destructive of these is the Atlantic hurricane, called a typhoon in the Pacific or a cyclone in the Indian Ocean. Great circular vortexes of air, they usually form near the equator over the ocean and then move slowly northward around an "eye," or a center of dead calm.

Although the eye of a hurricane is almost without wind, the air around it may reach speeds of more than one hundred miles an hour. (By definition, a hurricane is any wind of more than seventy-four miles an hour.) Most hurricanes in the Atlantic take place in the late summer and early fall. Why they form at this time of the year, why they form over equatorial ocean waters, and how they achieve such great energies are questions

A view of a cyclonic storm system located about 1200 miles north of Hawaii. This photo was taken on the Apollo 9 space flight in March 1969.

Courtesy National Aeronautics and Space Administration

still under study by meteorologists. The warm, moist air from the tropics is probably an important factor in their birth, but there may also be other influences.

Hurricanes generally lose their strength after they pass from water to land, but often not without doing great damage to shorelines and islands. For many years now, the United States Weather Service has detected, named and followed hurricanes. Each is given a woman's name—a name beginning with A for the first storm of a particular season, beginning with B for the second, and so on—and its speed and probable pathway are made public to alert the East Coast of the United States of possible danger. Although hurricanes usually move northward and vanish harmlessly at sea, their courses are unpredictable, and they have come ashore anywhere from Mexico to New England.

Typhoons and cyclones, though equally dangerous, less often reach land. However, they have caused great destruction in the islands of the Pacific and along the coasts of India and Africa.

Despite their immense power, hurricanes are not the most powerful winds in the world. The tornado, a circular vortex of wind which usually appears over land, may move at greater speeds. Tornadoes, however, are seldom as large or as long-lasting as hurricanes. Usually they form over the flat plains of the central United States, appearing in the sky as a dark, menacing circular cloud whose center may dip down and touch the earth.

A tornado is a whirlpool of air that becomes visible because the thousands of droplets of water within it condense as warm and cold air meet. When its vortex touches earth, it also picks up dust and dirt, which makes it even darker. Tornadoes are most common in spring and early summer, probably because at that time of year, deep layers of dry air rise over the Rocky Mountains and cover warm, moist air from the Gulf of Mexico.

The great speed of the tornado's wind is its chief method of destruction, a speed so great it rips buildings to bits, moves automobiles through the air, tears up trees, and generally leaves a wholesale path of destruction in its wake. But tornadoes also

A tornado cloud, or funnel, about to "touch down." The cloud is typical of a tornado-bearing part of the atmosphere—a thick, dark cloud low to the ground with a funnel emerging from its lower edge. Funnels often approach the earth like this without touching down and without causing destruction.

Courtesy National Oceanic and Atmospheric Agency

have a great difference in air pressure between their centers and the surrounding air. This acts as a kind of suction, causing buildings literally to explode as the tornado passes over them.

Tornadoes are more difficult to chart than hurricanes, simply because they appear and disappear with great rapidity, but the U.S. Weather Service does try to issue warnings of their probable approach during the months when they are likely to appear. The service also issues warnings of severe thunderstorms, another peril of the Plains states. Thunderstorms are created by the collision of warm and cold air. They may tower thousands of feet into the sky, sometimes almost to the limit of the troposphere. Because they do, they are a potential hazard to aircraft as well as to persons on the ground. A few aircraft have suc-

The funnel of a tornado after it has touched the earth, showing the spiral of destructive power as it races across the land. The dark cloud characteristic of this kind of weather formation is at right.
Courtesy National Oceanic and Atmospheric Agency

cessfully flown through severe thunderstorms, but pilots find it more prudent to avoid them whenever possible. Thunderstorms can cause severe damage to crops, especially when they precipitate hail, and the lightning that often accompanies them also is a potential danger to trees, buildings, and humans.

Mark Twain once wrote, "Everyone talks about the weather, but no one does anything about it." As with all changes in weather, man knows more about the results of violent movements of the atmosphere than he does about how to control them. He always has been at the mercy of the winds. Only in the past twenty years has he been even somewhat successful in weather modification.

Most of the efforts at weather control have been directed at wringing the moisture from the clouds through cloud seeding. Cloud seeding usually involves injecting clouds with substances that will cause their moisture to be precipitated. Both dry ice (frozen carbon dioxide) and silver iodide have been used. In either case, when "seeds" are sprayed into a cloud, they form nuclei around which water drops can collect to form rain. Seeding has been carried out both by dropping or spraying "seeds" from aircraft and by blowing the material into clouds from mountaintops. Not all meteorologists are convinced that cloud seeding is effective in producing more rain than would normally fall. Certainly it can wring no more moisture from clouds than they contain, and it is of no value in attempting to make rain fall from clear air. Real rain clouds must be present before cloud seeding can be effective.

Although other attempts have been made at altering the weather, the tremendous amounts of energy present in very violent forms of weather—hurricanes, thunderstorms, and tornadoes—make it unlikely they can be brought within man's control for a long time. Instead, a more practical approach seems to be the gathering of more and more knowledge about the forces which control weather naturally. This kind of information can help in charting and warning of violent weather changes and can decrease the loss of life and damage to property which such events in the atmosphere now cause.

Although much is known about the movement of the air, then, much remains to be discovered. The air is still filled with mysteries. Not the least of these is how man is to continue to survive within the invisible cloak of the atmosphere, even as he continues to abuse the vital gases which gave him life. To learn these secrets before permanent damage makes the atmosphere uninhabitable will require much more research and understanding than man now possesses, but it is a goal toward which much scientific dedication is now directed.

—3—
FIRE

More than three centuries ago, the Incas of ancient Peru built temples high in their mountain cities in which to worship the sun. They were one of many ancient civilizations which considered the sun divine. Even then, men were aware of the importance of our star to life.

Today, we no longer consider the sun a god, yet we retain an immense appreciation of its power and value to all living things in the biosphere. Without the sun, no life on earth could survive. Its energy, caught and stored by green plants, is basic to all our food. Sunlight powers the winds and waters of the world. It is the driving force of the biosphere.

Moreover, the sun does its work with a remarkably steady heat and light. Its rate of burning is so stable that the average temperature of the earth varies only slightly from year to year. Even in periods of intense cold, as during the ice ages, the sun continued to nourish the earth.

The sun's energy arrives at the earth's surface in different forms, all a part of the electromagnetic spectrum. The electromagnetic spectrum includes not only visible light, light we can see with our eyes, but potent invisible rays of energy as well.

Radiation from the sun, of both higher and lower frequencies than that in the "visible window" of light radiation, showers on the earth all the time. Although visible energy can be seen with the eye, most energy is detectable only by a variety of different measuring machines invented by man. Solar radiation varies greatly. X rays make up a part of the energy poured out by the sun every day. The sun also emits atoms of its principal gases, hydrogen and helium, sending them as ions across the millions of miles of space to the earth.

This great aggregation of particles and rays flows not only to our planet, but to all the rest of the planets of the solar system. Although only a small part of it ever reaches the earth's surface, perhaps a billionth of the sun's daily total output, this is enough to make the earth's surface a place where life can thrive.

The sun's emissions are called the solar wind. How far the solar wind extends into space is not certain because the outer limits of the solar system have not yet been sampled by space probes, but probably vestiges of it are to be found around Pluto at the edge of the nine planets in the sun's gravitational field.

The earth plows through the solar wind somewhat like a missile moving through the atmosphere. As it pushes through the solar plasma, it creates a bow or shock wave before it, while an extended tail, perhaps 80,000 to 100,000 miles long, trails behind it. The earth's magnetic field also has an effect on the particles of the solar wind, deflecting them upward over the earth and allowing them to be fed into this area of space.

The existence of the solar wind was not discovered until the first satellites were flown in orbit around the earth in the 1960's. Then satellite samples indicated not only the presence of the solar wind, but the ability of the earth's magnetic field to deflect it. The belts of magnetic force are now called the Van Allen Belts, after the scientist, Dr. James Van Allen, who developed the space probe that located them.

The Van Allen Belts are shaped somewhat like doughnuts, thick and deep at the earth's equator and shallow or nonexistent at the poles. At their deepest, they extend into space for several thousand miles. Without them, the earth's surface would be

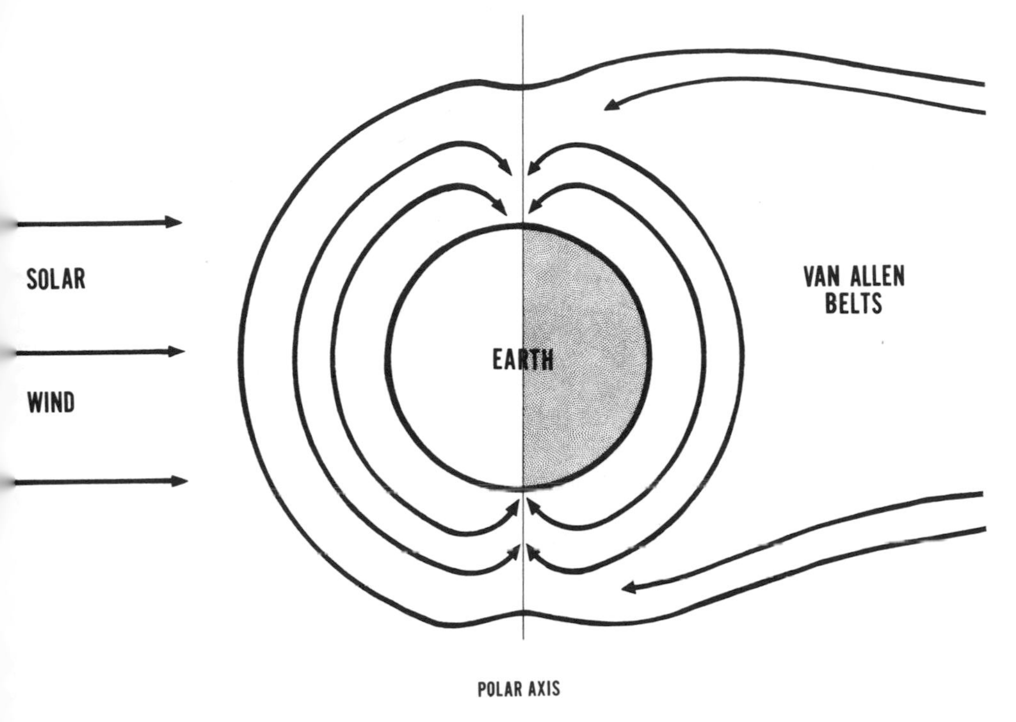

The pattern of the earth's passage through the solar wind, showing the location of the Van Allen Belts, the bow, or shock, wave preceding the earth through space, and a portion of the "tail" of particles of the earth's magnetosphere.

exposed to much more high-energy radiation from the sun than it is now.

The Van Allen Belts, however, do not trap all the sun's energy. The ozone layer also helps to screen out ultraviolet rays, energy at the lower end of the visible part of the spectrum. Because ozone is composed of ionized oxygen, it is possible that the ozone has not always been present around the earth. This can be inferred from the fact that it may vary in depth during certain times of the year. No one knows what the earth was or would be like without ozone, but many scientists have speculated that life on earth would be affected by its disappearance, perhaps adversely. For example, without its presence to shield life from ultraviolet radiation, there might be an increase in radiation-caused cancers.

It is because of this possibility that these same scientists recently have become concerned about the increasing use of aerosol spray cans. A variety of substances now are marketed in such cans, ranging from hair spray to paint. In all cases, the material in the cans is sprayed into the air because of the pressure of inert gases added to them. The gases do not combine with the useful materials in the cans, nor do they combine with most of the gases in the atmosphere, but they may have the effect, if released into the air in sufficient quantities, of blocking or destroying the ozone.

While not all scientists agree this is likely to happen, all concede the loss of the ozone's protection would be a serious matter for living organisms. The question seems to revolve not around this fact, but around the question of how much inert gas in the atmosphere is too much. Most scientists want additional studies of the problem, to make clearer the dangers involved.

Even if the ozone were decreased or destroyed completely, the atmosphere itself would still serve as some protection from the powerful radiation which the earth receives from the sun, for some of it is screened out by the atmosphere itself. Radiation also loses some of its energy as it is bounced back and forth from the earth's surface to the molecules of the heavier parts of the air.

Further, the earth does not receive equal amounts of radiation over its entire surface, because of an effect called insolation. Insolation takes place because the earth is a sphere instead of a flat surface. Because this is so, sunlight strikes most directly at the earth's equator. Near the poles, sunlight not only must travel farther through the atmosphere to reach the earth, it also approaches the surface at a greater angle. Thus, the equator receives more direct radiation than do the poles. The effect is further enhanced because the earth is not perfectly perpendicular in relation to the sun. It is tilted on its axis at an angle of 23.5 degrees, and this allows some parts of the world to receive greater sunlight at certain times of the year than at others. The North Pole, for example, is completely without sun at certain times of the earth's passage around the sun. So is the South Pole. The temperate parts of the earth receive varying amounts of sunlight during certain times of the year because of their changing relationship to the sun as the earth completes its orbit around our star.

Nevertheless, the earth's overall exposure to solar energy is remarkably constant and predictable, so much so that scientists have been able to calculate with considerable accuracy the amount of energy it receives over its surface each day. An equation taking into account the amount of sunlight deflected by the atmosphere, its clouds, and the ozone, the amount absorbed by the ocean's waters, and the amount released by the seas, shows that only about one quarter of the one billionth of the sun's daily total output of energy finally penetrates to the earth's surface.

The amount of sunlight reflected back into the atmosphere after it reaches the surface is also variable, depending upon the surface. Most of what reaches the ocean is absorbed—even though it may later be released into the air—but as much as 90 percent of the sunlight that strikes the ice- and snow-coated surfaces of Greenland and Iceland bounces back into the atmosphere.

Even though diluted by all these various effects, the amount of solar energy far exceeds anything man has been able to turn into heat from the earth. It is this immense energy output that runs the earth's biospheric engine, creating winds, rain, fog, and snow, moving water from place to place through the air over both oceans and land. Were it not for the sun playing on the atmosphere, evaporating the ocean's waters, moving them with wind over the land, and causing their condensation into precipitation, there would be no way for living plants to flourish on the continents. The earth would be a desert.

The sun's heat is also necessary for the many complicated biochemical reactions which are a part of metabolism, the interactions by which living organisms are able to extract energy from biological materials. These interactions are dependent on the ability of the earth's green plants to "fix" or store energy. If all the heat received from the sun each day had to be used at once, the earth would be energy poor. Fortunately energy can be incorporated into biochemical molecules and stored.

The simplest example of such storage is a tree. Trees are green plants, and it is in their leaves that sunlight is first captured. There, minute structures called chloroplasts turn the sun's radiant energy into stored potential energy, as a part of photosynthesis. Just as other plants also convert radiant energy into stored energy, trees can hold energy within them until either they are burned or are consumed as food.

Photosynthesis is dependent on the element carbon, one of the most abundant on earth. Carbon is an element capable of arranging itself in long chains, in rings, or in other structures. In such configurations, carbon molecules can be made into many different substances by the addition or subtraction of only a few atoms.

The most obvious way to release the energy which a tree contains is by burning it for heat, but decay also accomplishes the same thing much more slowly.

The burning of a tree or the use of a plant for food does not destroy the energy it contains, but it does transmit it to an increasingly limited number of useful places. As energy is used, it becomes more and more disassociated from itself and more and more randomly distributed. This helps to explain why heat energy always seeks a cooler place to which to move after it is released from its original "package." As energy is more and more divided, it becomes so widely separated that it is no longer useful. Presumably, at some time in the future, all energy in the universe will be so randomly dissipated into space that the universe will have run down and will, for all practical purposes, be dead.

The question of the ultimate fate of the universe and the energy it contains is, however, beyond the scope of this book. What is more important for us is the fate of energy in the biosphere. The biosphere contains a large amount of energy.

Moreover, it receives large additional amounts from the sun each day. Unfortunately for human purposes, it takes much longer to fix easily obtained sources of energy in the biosphere than it does to use them up.

Wood is an example of fixed solar energy that is easily destroyed. Many of the earth's forests have been burned for the fuel they contain. Oil and coal are also fixed energy sources. Both these fuels were formed from plants and animals compressed within the earth's surface over millions of years. (It is because they are formed from the dead remains of once living creatures and organisms that they are often called fossil fuels.)

Unfortunately, it takes only weeks or months to receive, refine, and consume fossil fuels, and mankind is presently using them at such a rate that they will soon be exhausted.

Man also is rapidly using up as food the solar energy fixed in plants and animals. The energy in living organisms is dependent on the food chains of the earth. Food chains begin with primary energy gatherers, such as the plant plankton of the sea. These primary energy gatherers are eaten by a succession of larger and larger creatures until the creatures at the top of the pyramid of species is reached. In the sea, for example, the end of the chain begun with plankton may be whales. The effect of food chains is to concentrate more and more energy in larger and larger creatures, limited by their size and energy requirements to smaller numbers.

Man surmounts all food chains. He is not a primary energy gatherer, but he is the paramount and primary consumer of all sources of energy, and either interrupts or is at the top of all food-chain pyramids. Man has always been dependent on other species for survival, but until the past half dozen centuries, he has always been able to live in balance with other food chains. Today his influence on many has become disastrous, not so much because he consumes all kinds of energy-bearing organisms, as because of his increase in numbers.

During the past three centuries, the number of human beings on the earth has doubled and redoubled at an ever-accelerating rate. One scientist has likened this effect to that of a virus spreading swiftly across the globe, living on the remainder of earth's creatures, much as a disease virus steals nourishment from the cell it invades until it kills it. The situation has become so critical that man's own ability to survive has come into doubt.

Agriculture, the ability of human beings to increase the amount of energy-giving plants and animals through cultivation, has been blamed for both the increase in numbers and the loss of energy sources by some scientists. Man has become increasingly dependent on agriculture for survival since the end of the last ice age.

Planting domesticated crops increases the amount of food (and energy) available from any given area of the earth's surface, but it also requires that an increased amount of energy be expended to reap the harvest.

Consider the reduction of the earth's forests. Ten centuries ago, early farmers found they could increase crop production by cutting down and burning forest trees, and by then planting the cleared land, a system called slash and burn. But they also discovered that within a few years the land lost its natural soil nutrients. Without the addition of fertilizer, crops declined.

Farmers could move to another part of the forest and repeat the process, but this practice was limited by the amount of land available. So, in time, fertilizer became the accepted way to increase production. At first, fertilizer came from human and animal wastes, spread on fields as manure. Because they contained nitrogen and other plant nutrients, they returned to the soil the necessary elements to make plants grow. Later these same elements, notably nitrogen and phosphates, were extracted by chemical processes and made into commerical fertilizers. Commercial fertilizers could not be produced, however, without the use of energy. The use of fertilizers to increase crop production thus has been made at the cost of an additional expenditure of energy.

Early farmers also soon discovered crops could be made to grow more readily in some parts of the world if the land was irrigated with water. At first, natural runoff waters were channeled to dry lands or allowed to flood them, but men rapidly realized that canals and aqueducts were needed if much dry land was to be used for crop production. Dams, canals, and other waterways rapidly came into use, even in early civilizations. These, too, required the use of energy for their construction, and the cost of such energy had to be added to the cost of production.

The increased storage of solar energy in larger agricultural production thus must be measured against the additional expenditure of energy required to grow it. By subtracting the amount of energy needed to grow crops from the amount they are able to provide in additional food production and energy for humans, it is possible to get a net balance. The net balance of energy provided, not the total amount of food produced, is what really matters in examining increased agricultural production.

Experts who have examined such figures believe, despite the use of more and more of the earth's surface to grow food, that the balance of energy available in the world remains about the same today as it was before agriculture began. This might not be so serious a problem were it not for the fact that the numbers of human consumers of food have grown manyfold. Thus, even though food production has increased, it has been insufficient to feed the larger and larger human population of the earth.

In addition, the increasing use of the earth's surface for agriculture has disturbed many of the delicately balanced food chains naturally present on our planet. As the great forest which once covered large parts of North America and Europe has been removed to make new farmland, much additional natural life has disappeared. Once adequate to feed men who were primarily hunters, the new agricultural land has made human beings ever more dependent on finding new and more productive ways to grow food.

The situation has grown more acute in the less-developed parts of the world, principally Asia, South America, and Africa. In the search for new land areas to plant, hillside forests are being further reduced. In jungle and tropical rain-forest areas, slash-and-burn agriculture has forever erased lands capable of maintaining natural plant life, but incapable of growing domesticated crops.

Where hillside forests are cut down, rain washes away the topsoil because no tree roots are available to keep it in place, and large parts of mountains have become sterile deserts. As the need for food continues, the process accelerates. Men climb higher and higher, remove more and more forests, and wreck forever larger and larger places on mountainsides.

The reduction of natural plant life has another, even longer-range effect on the ability of the earth to capture the power of the sun. Millions of years ago, when the forests and swamps from which our present coal and oil were made grew, the natural creation of these two precious fuels took place with ease.

The destruction of today's forests will make unlikely the creation of new fossil fuels. The natural materials needed to create them will be gone. At the same time, the consumption of existing coal and oil reserves is moving at such a rapid rate that our present known supplies of these two fuels soon will be exhausted.

At first it was possible to find both lying at or near the earth's surface, but as the easier-to-reach pools and seams of oil and coal were uncovered and used, men had to dig deeper and deeper to find new supplies. Coal still exists in seams near the surface and can be extracted by strip mining—that is, by following a vein of coal across the land, no matter where it goes. Strip mining often has the same effect on the land as the destruction of natural forests. Unless reforested, the remaining topsoil is washed away, leaving infertile, barren land on which little or nothing will grow.

Most shallow pools of oil already have been drilled, and many of them have been pumped dry. Oil explorers have been forced to dig deeper and deeper into the earth or to search in remote parts of the world—the Arabian peninsula, the North Slope of Alaska, or offshore along the continents. Probably reserves of oil still exist beneath the continents, but at depths which make it more and more expensive to reach them.

The consumption of oil to power machinery has accelerated year by year. Many experts believe the world already has passed through the peak of oil production. Despite the discovery of new reserves of oil—in Alaska or off the continental shores of the United States, for example—oil seems destined to be in ever-decreasing supply in the years to come. Coal is still abundant in some parts of the world, notably the United States, but it, too, eventually will be less and less available. As oil disappears, greater dependency on coal will also increase its consumption.

Nuclear energy also is a potential source of power. By refining from the natural radioactive ores of the earth's crust and fissioning it in a nuclear reactor, natural heat can be used to heat water and such elements as sodium to make steam. Steam then can be used to turn electrical generators. Nuclear ores only recently have been tapped for the energy they contain, and a large supply is still available to use. Moreover, some nuclear reactors have the ability to "breed" additional supplies of fissionable material. In a breeder reactor, not only does the reactor's core supply heat for the creation of steam, but nuclear fission also creates additional supplies of radioactive plutonium. Plutonium is not a naturally occurring element in the earth's crust, but it is a by-product of the fission of radioactive uranium. Unfortunately, plutonium is a highly poisonous substance and retains its poisonous characteristics for many thousands of years. Only a tiny amount of it is sufficient to do great damage to living organisms. How much and how successfully it will be used for the energy needs of the years ahead is uncertain.

The direct power of the sun remains, however. It is available for energy in two ways, directly as it reaches the earth from space and as a model for the fusion of hydrogen atoms in the thermonuclear process. Thermonuclear energy is the basis for the explosion of a hydrogen bomb. By surrounding the uranium or plutonium core or trigger of a bomb with hydrogen, it is possible to use the great heat and pressure created by the trigger's explosion to force hydrogen atoms to duplicate the process they undergo in the sun's core. The reaction lasts far less than a second, however, and is not controlled. It simply explodes with immense force.

If, however, it were possible to control the energy released from the fusion of hydrogen atoms gradually, almost unlimited amounts of power would be available to man. The seas are rich in hydrogen. By using them to obtain fuel, it would be centuries, perhaps thousands of years, before the earth's supply were completely converted to helium.

Yet while this idea is simple enough to state, it has proved far more difficult to make a reality. The difficulty lies in the tremendous amounts of heat generated by the thermonuclear reaction. No container on earth is strong enough to withstand such power. It is, however, possible to construct "bottles" of magnetic lines of force which will hold fusing atoms for a short time. Although present magnetic bottles tend to bend and buckle, allowing the gas within them to escape, it is probable that, if thermonuclear energy is to become useful, some perfection of such a system will be necessary.

Scientists have been working for more than twenty years to overcome the difficulties in the concept of thermonuclear power. No end to their efforts is yet in sight, but the possibility of its use remains.

Finally, direct sunlight is also a source of energy, perhaps the ultimate source to which mankind will have to turn as other forms of fuel are gradually consumed. Far more sunlight falls on the earth each day than is used, either by plants or by animals. Much of it is reradiated back into the atmosphere and bounces back and forth between sky and earth until it loses its power. If even a small portion can be captured as it strikes the earth's surface, it can be made to do useful work for mankind.

As with thermonuclear energy, however, reality is far away.

It is possible to capture sunlight directly by using photoelectric cells, which are able to convert light directly into electricity. Indeed, such solar panels already have been used as a power source on some spacecraft. To gain much power, however, a large area of photoelectric cells must be exposed to the direct rays of the sun. Cells are expensive to build and the cost of their construction at present is more than the amount of useful energy they can provide. Their use in spacecraft happens because the craft to which they are attached needs little electricity and because no other natural source than sunlight is available to a moving vehicle in space.

Further, some parts of the world do not receive sufficient sunlight to make them likely places where solar panels could be set up. They have long nights in the winter, making it impossible for much power to be generated at a time when it is most needed.

However, it is possible to design houses so that they accumulate heat during the day and use it during the night. By proper siting of the house—placing it in proper position in relation to the prevailing sunlight—and by allowing for heat traps and methods of air circulation, houses in warm, temperate climates can be lived in without any other kinds of energy at all for heat and light.

Almost any house, even a solar-designed house, however, needs some auxiliary power, usually electricity for those periods when the sun fails to shine. Engineers have already designed houses which are heated and cooled by heat pumps, simple devices which collect heat during the winter and exhaust heat to the air during the summer.

Present heat pumps usually are a collection of pipes filled with a refrigerant. As the refrigerant moves through the pipes, it absorbs heat from the air. The heat is removed from the refrigerant at a pump through a heat exchanger and blown into a forced-air heating system. The system can be made to work in reverse in summer months. As the heat in the air of the house passes through the heat exchanger, it is fed into the refrigerant and forced through the pipes. There it is exhausted to the air.

Present heat pumps are not efficient enough to be able to heat completely most present-day houses. They must have an auxiliary source of electric power for the times when they are unable to supply sufficient heat through their own system.

Finally, the effect of the sun as it drives the atmosphere also may someday be harnessed for useful energy. For it is the sun which makes wind, and wind has energy within it. The best known form of useful wind energy is the windmill. In the days before the widespread use of electrical energy to power pumps, windmills were used in many parts of the world to draw water, to grind grain, and for similar purposes. Most such windmills were constructed with large, light blades or with many small blades set at an angle to absorb the energy of the wind. Both in the Netherlands and in the western United States, much pumping was done with wind power, just as ships were pushed across the ocean by sails.

The great difficulty with wind as a source of solar energy is its variability. Winds tend to come and go, but in some parts of the world, prevailing winds are usually regular and of sufficient force to be useful as a power source. Their most likely use today is not to turn machinery directly, but to generate electricity, which then can be used to power engines. Electrical energy generated by turning windmills can be stored in batteries and released for use when the wind is not blowing.

Windmills also need not be traditional in shape. A variety of different kinds of windmills, which do not have the traditional kinds of blades, are now being studied, both in the United States and elsewhere. Many look much like airplane propellers, but some are peculiar ribbons of light material housed in rotors. The task of windmill designers is to increase the efficiency with which the mills turn. To put this another way, the task of the windmill maker is to reduce the friction with which the windmill operates so that as much as possible of the energy contained in moving air is transmitted directly to the electrical generator attached to the device. Windmills have the advantage of almost unlimited space in which to be mounted above houses or factories. However, they do not seem adaptable for places where there is no strong prevailing wind.

Along seacoasts, in plains areas, and on mountaintops, however, where a steady, regular wind blows, windmills may be a valuable addition to efforts to harness the power of the sun for useful work.

Probably there are other methods of achieving power and energy as well. The burning of garbage, the use of methane gas from sewage and garbage, and many other potential sources of energy are now being explored in different parts of the world. Likewise, much effort is being spent in attempting to convert the energy still available from the biosphere, for the wasting of energy is as serious a problem as the eventual exhaustion of the energy still available.

The biosphere remains the place where energy can be put to useful work. It also is still the place where much energy is still available. If the biosphere is to remain what its name implies, a sphere containing life, it must have such power. Without energy, life cannot survive for long, at least human life as it is now known.

To understand this truth, man need no longer worship the sun, but he can never cease to appreciate it.

—4—
WATER

The earth's second great ocean is the sea. It stretches over three fourths of the world's surface and is our planet's most obvious characteristic, one which distinguishes it from all other planets in the solar system.

As most schoolchildren quickly learn, the sea is made of two gases, hydrogen and oxygen, combined as a molecule. The sea is also salty. It is its salt taste that differentiates it from the fresh waters of the continents. Salt is a molecule whose elements are sodium and chlorine, but sodium and chlorine are only two of many elements dissolved in seawater. At least forty others have been detected, although some of them are present only in minute amounts. Moreover, the salt content of the sea varies from place to place. It is greatest in the Red Sea between Africa and Arabia, but less in the earth's polar regions.

The depth of the ocean also varies from place to place and has been higher or lower from time to time in the earth's past, depending on whether or not the earth's climate has been colder or warmer. Today, shallow seas exist in places where there once was no ocean at all, whereas in others, deep trenches make the ocean miles deep. In the western Pacific off the Philippine

Islands, for example, as much as eight miles of water may lie beneath the surface. On an average, however, the ocean is about 12,000 feet deep, a distance as great as the altitude of many of the mountain ranges on the land surface of the earth.

The sea is opaque. It is impossible to see its bottom because the ocean's waters reflect light. Sunlight does penetrate a few hundred feet, a fact of vital importance to all living things on earth. For it is these few hundred feet, spread over the ocean's surface, that are the pastures of the sea, the places where plankton and diatoms, tiny marine plants and animals so small they can be seen only with a microscope, live by the billions. The planktonic pastures are the beginning of many of the food chains of the earth. Without them, the pyramids of interlocking relationships by which one species feeds upon another would not exist. In addition, plant plankton are an important source of oxygen and a consumer of carbon dioxide, and they play a key role in the carbon dioxide–oxygen cycle which sustains life in the biosphere.

Plant forms of plankton, like the green plants of the continents, capture sunlight as it falls on the sea and store it so that it can later be converted into food energy by other life forms.

Below the pastures of the sea, life is less abundant. Yet it exists at all levels of the ocean, even in the inky blackness at the bottom. Exploration of the deepest part of the sea, the bottom of the Mariana Trench in the Pacific, has turned up sea worms, starfish, and other creatures who live by eating animals from layers above them.

Life, however, is not uniform in the ocean. It is most abundant near the equator, and becomes less as one approaches the poles.

Like the air, the sea is never static. It never remains still, and its waters, both on the surface and in its depths, are constantly moving. Three forces work to drive it—currents, tides, and waves. Each, to a greater or lesser degree, affects the other. Although all the reasons for the movement of the sea are not completely understood, ultimately it probably is caused by forces beyond the earth.

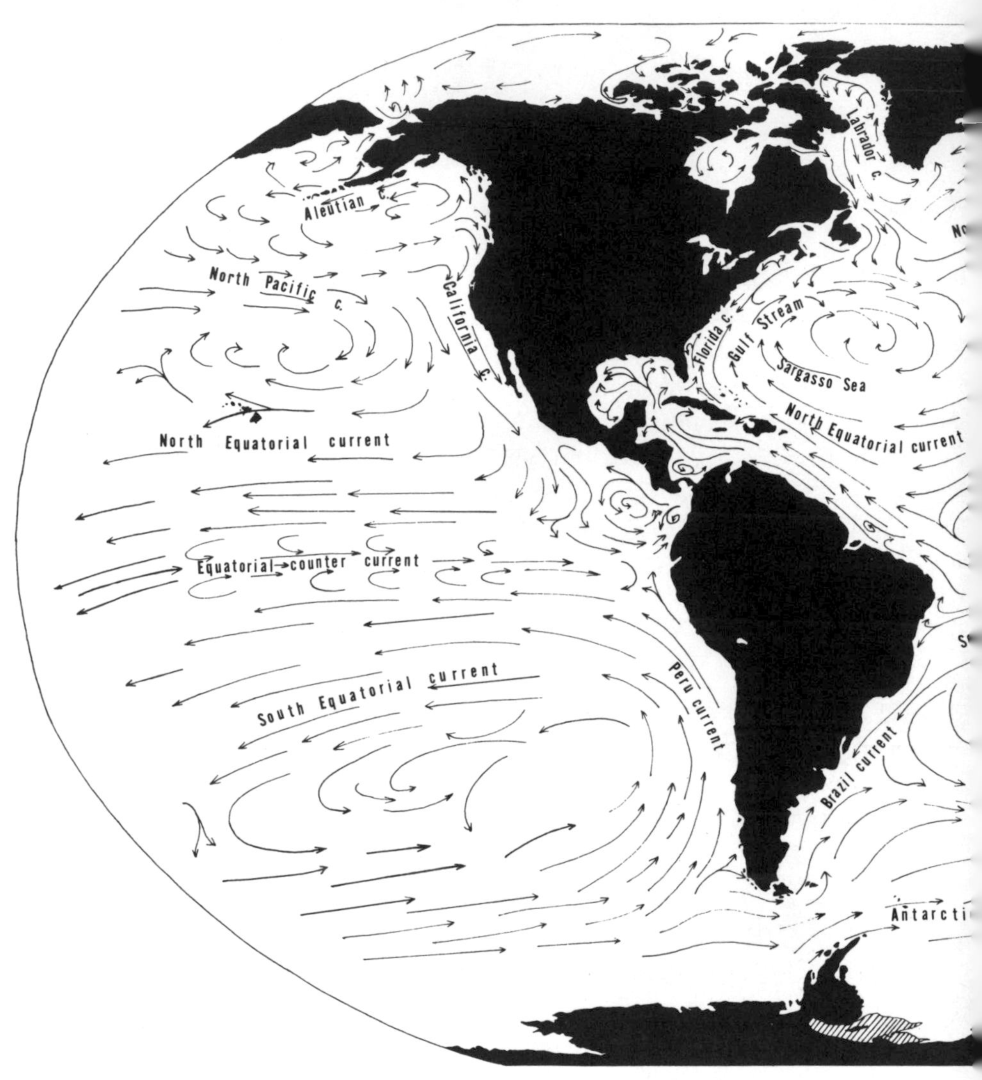

The principal currents of the ocean in both the Eastern and Western Hemispheres of the earth. Lesser currents are not outlined.

The rotation of the earth, its movement in relation to the moon and the sun, and the radiant heat of our star all have their effects on the sea, causing it never to be at rest.

Ocean currents have been studied for many years, and oceanographers now know the patterns by which the sea moves. Currents usually flow in opposing directions on opposite sides of the Equator, but their movement also may be affected by the shape of the shorelines and the presence of islands.

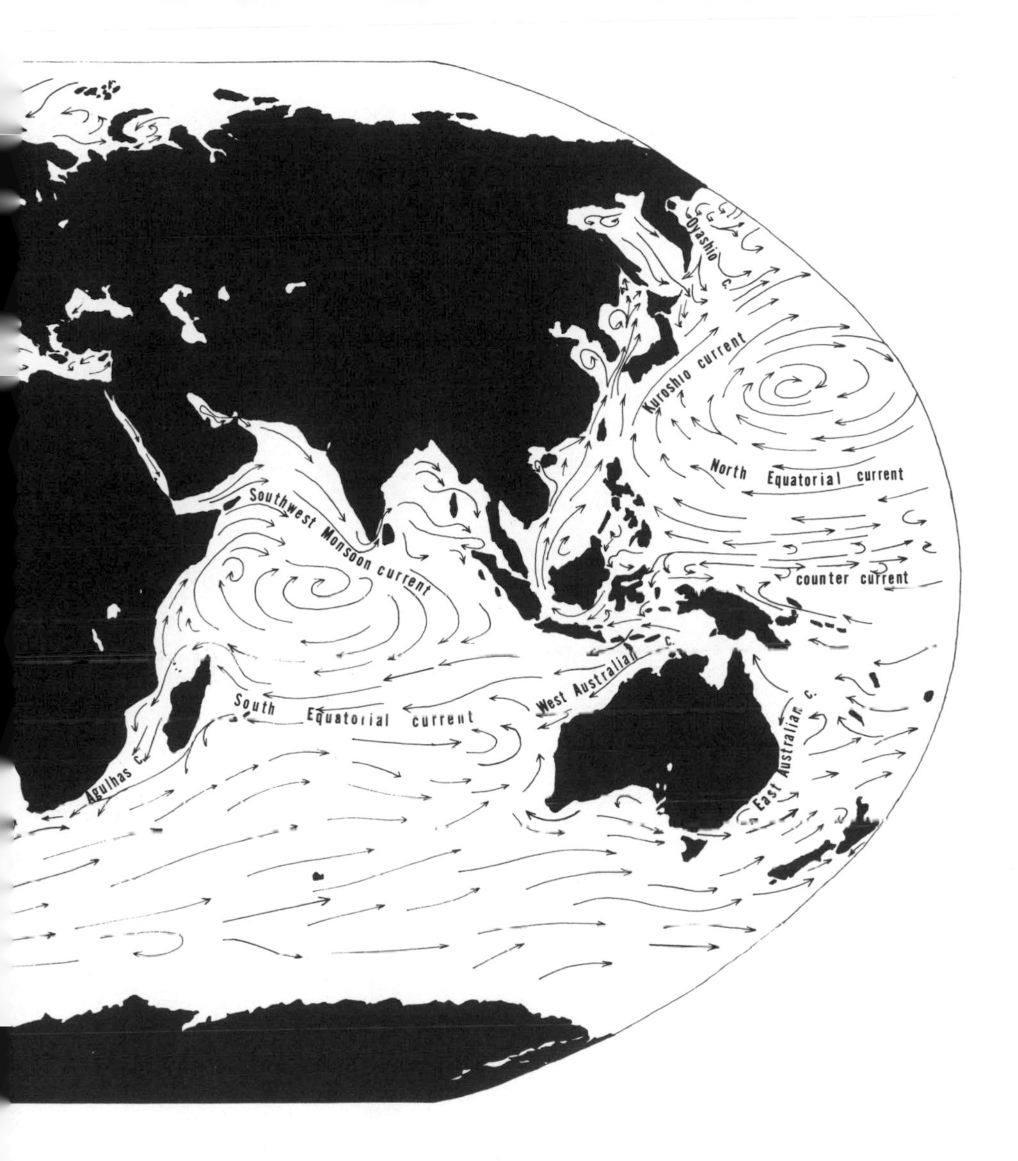

The longest sweep of the sea's currents is along the equator. North of the equator in the Atlantic is the North Atlantic Equatorial Current, which flows westward from near Africa across the Atlantic to South America. There it is turned northward by the shape of South America and becomes the Gulf Stream, perhaps the best known of ocean currents. The Gulf Stream, named for the Gulf of Mexico, is not a single current but a number of fingers of warm, north-moving water which flow along the Atlantic Coast of the United States toward the Arctic.

Before the Gulf Stream reaches Arctic waters, however, it is diverted by the colder waters of the Arctic Ocean and flows past Greenland and Iceland, descending along the coast of Europe to reach its beginning point off Africa.

A similar equatorial current south of the equator flows in the opposite direction, eastward from South America toward Africa, but it is cut off by the shape of the continent of Africa and does not reach as far south as the Gulf Stream pushes north toward the North Pole.

In the center of the North Atlantic, out of the Gulf Stream, the ocean's waters are relatively stagnant. Circling slowly about, they form the Sargasso Sea, in which a vast collection of kelp and other aquatic plants live seemingly little affected by the movement of the ocean. The Sargasso Sea was once thought to be a place where hundreds of sailing ships became becalmed derelicts, unable to escape its waters, but this is a myth.

North and South Equatorial Currents also exist in the Pacific, but their flow is much less disrupted than in the Atlantic. For almost half the earth's circumference there, the equatorial currents flow eastward and westward unrestricted. The North Pacific Equatorial Current eventually is deflected by the islands of the western Pacific—the Philippines and Japan, principally—and swings northward toward the Aleutian Islands and Alaska. Then it returns down the west coast of North America as the Japanese Current.

The South Pacific Equatorial Current flows along the equator toward South America, travels down its coast, and then is caught up and deflected by the Antarctic Circumpolar Current. The latter is the swiftest and strongest current in the world. Its cold waters swirl around the ice-covered southernmost continent at great speed, unimpeded by any islands. As it passes between South America and Antarctica—where only six hundred miles separate the two continents—it can move as much as 3.5 billion cubic feet of water past a given point in a single second.

Coincidentally, the seas around Antarctica are the stormiest in the world. Extreme winds and waves follow the path of the current and make it one of the most hazardous parts of the ocean in which to navigate. In the days when sailing ships had to pass through the Strait of Magellan or around Cape Horn to reach the Pacific, it often took many days to force a passage through the swiftly moving current.

Oceanographers recently have come to believe that the oceans and their currents play a much more important part in the weather of the biosphere than they once thought. Although it has long been assumed by oceanographic scientists that the ocean's currents are generated by the wind, some new studies of the deep sea make it seem more likely that they result from differences in water temperature and the amount of salt in different parts of the world's ocean.

Not only may the deep ocean currents result from these forces, they also appear to be much more complex than earlier studies have indicated. Instead of simple surface and deep currents and countercurrents, the oceans now seem to be layers of differing kinds of water moving in directions opposite to one another. In some cases, these undersea currents may be thin. In others, they may be large pools or eddies, somewhat like the highs and lows in the air. The eddies of the ocean are probably not as large as those in the air, but they may be of major importance in affecting the air above the surface of the sea.

The size and variability of eddies make it difficult to chart their movements. They may change temperature and location very rapidly, often within a matter of hours or days. The situation becomes even more complicated because of the ability or lack of ability of the sun to heat the waters of the sea in the various parts of the ocean. If these factors all truly affect the movement of ocean waters, then it is possible to study deep-sea eddies only by both measuring the amount of sunlight falling on the oceans and by examining underwater layers over large parts of the sea. To do the latter requires sinking measuring devices and holding them at the proper depths for certain periods of time. The information such probes gather also must be relayed to the surface rapidly and accurately. Because the sea is so vast, this becomes a task even more difficult than making daily measurements of the differences in air pressure and air temperature, a task for which no nation, or even a group of nations, is yet prepared. Research toward a better charting of ocean water changes has begun, however. In 1974, a group of forty ships from many different countries converged in the Atlantic Ocean to study the effects of the ocean on weather. The group was called the Global Atmosphere Research Program (GARP), and the project itself was termed GATE (GARP's Atlantic Tropical Experiment).

The fleet of oceanographic vessels made measurements of both air and water temperatures and movements over a wide area of the ocean. In addition, weather stations in fifty-seven countries also took part in the experiments. One of the experiments included deep-sea measurements of water movement using Swallow floats.

These devices, invented by Dr. John C. Swallow, an English oceanographer, are anchored at different levels in the sea to measure temperature and to chart the movement of eddies. Each consists of a sound transmitter in an aluminum tube sealed at both ends. Because aluminum is slightly less compressible than seawater, the tubes sink until their density equals that of surrounding seawater. It also is possible to locate them at other depths by weighting them properly. The probes send continuous sound signals to listening oceanographic vessels on the surface, recording the speed with which the water at a particular level is moving.

The information from GATE Swallow probes, properly processed by computers, showed there were eddies indeed in the sea. They were less specific about whether or not such eddies are related to changes in the air and hence to changes in the weather.

Experiments aimed at relating changes in seawater temperature and movement with changes in the weather are continuing. One possible effect of such "highs" and "lows" in ocean water is the changes they may make on the regular surface currents, such as the Gulf Stream.

It is possible that below-surface eddies of water may cause sudden heating and cooling of the air above them. Eddies also from time to time may affect the regular movement of currents like the Gulf Stream, but it is probable that the relationship between the sea and the sky is more complex than this. Very likely, the world's weather machine operates because of a combination of even more factors than these.

Some oceanographers, for example, have suggested that the cold water of the polar seasons, the Arctic Ocean, and the waters which flow around Antarctica affect not only the movement of ocean waters, but also the ocean of air above them. Because heat always seeks a cooler surface to which to flow, the movement of both air and water is affected by the cold of the North and South Poles.

Very little is known about the polar areas of the globe, because until very recently year-round exploration has been extremely difficult. In the past half century, however, observation posts have been established on ice islands in the Arctic Sea, and permanent stations have been located in Antarctica. The Arctic ice islands have been manned for a year or two, enough time to permit them to follow the circular pattern of ice moving around the Arctic Sea between North America and Asia. Antarctic stations have been established on a permanent basis by many nations, not only along the outer edge of the frozen continent, but even at the South Pole.

From such studies, it now seems certain the movement of the jet stream in the Northern Hemisphere in relation to the earth's surface is the chief indicator of changes in the weather for this part of the world. Although there may be no "normal" pattern for weather, the wintertime variation of the jet stream's movement seems to give a good indication of how severe or mild the winter will be. The summer of 1972, for example, was severe in many parts of the Northern Hemisphere and was marked by a dislocation of the jet stream northward, particularly over the Soviet Union, causing a failure of that nation's wheat crop. Over the United States, however, the jet stream dipped southward, causing heavier than usual summer rains, again with serious effects on the American corn crop. (A hurricane off the east coast of the United States that summer added to the world's problems with weather.) The relocation of the jet stream seems to have come about because of differences in the temperature of ocean waters in the Atlantic. The water was very cold off Greenland, colder than usual for the summer. What caused the cold water? The answer is not known. It may signal a change in the general temperature level of the Northern Hemisphere. Perhaps it was only a local variation in the weather.

Local differences in the weather happen every year in some part of the world, but it remains a question as to whether or not they are related to more general changes in the world's weather. Certain local weather changes probably are unpredictable.

Perhaps the most predictable of all changes in the sea are the tides.

Tides usually are visible only along shorelines, either of continents or of islands. They appear to raise and lower the level of the sea, but this is not how they work, nor are tides the same height everywhere in the world.

Tides are caused by the interaction of the earth, the moon, and the sun. Were the moon not revolving around the earth, there would be no tides, and the ocean, except for waves, would lie flat on the earth's crust. As the moon revolves around the earth, however, it exerts a gravitational force, causing the oceans to bulge outward "under" it. A bulge also develops on the opposite side of the earth from this point because of the centrifugal force with which the earth is turning on its own axis. These two bulges in the ocean create tides on opposing sides of the earth. Both such tides are "high" tides, while "low" tides result at points on the earth 90 degrees from the bulges.

When the sun is in proper position in relation to the earth and the moon, it also exerts a gravitational force and has its effect on the tides. In the first instance, when the moon is "new" and most of its surface turned toward the earth is not lighted by the sun, *spring* tides are created. The word "spring" comes from an ancient Saxon word, "sprungen," meaning "full" or "strong." In the second case, when the moon is in its last quarter, or half-lighted by the sun (as a view on earth), *neap* tides result. "Neap" is believed to be derived from an old Scandinavian word meaning "barely touching" or "hardly enough."

Tides do not always arrive in the same place, however, because the earth is turning on its axis as the moon proceeds around the earth. Thus, at some times the moon is over a different part of the earth than it was when the tide was first created. The moon also appears fifty minutes later every day in its passage around the earth and delays the arrival of the next tide by twenty-five minutes. The moon also does not pass directly over the earth's equator in its orbit. Rather, it moves in an ellipse. This causes a variation in tides. One type, the semi-diurnal or semi-daily tide, arrives every twelve hours and twenty-five minutes. The other kind, a diurnal tide, comes only once every twenty-four hours and twenty-five minutes.

A further complication is that the earth is tilted on its axis with respect to the sun. This produces two high and two low tides with minor differences in height in some parts of the world, particularly the east coast of the United States and the west coast of Europe. In other parts of the world, however, there may be only a single high and low tide each twenty-four hours. In still other parts of the earth, there are two high and two low tides each day with considerable differences in height.

All this information is only a general description of the tides, however. The shape of the ocean basin in which they occur, even the shape of individual bays and inlets, has a considerable local effect on the height of any tide. This causes water in the local area as it is affected by the tides to move back and forth or to oscillate—just as light and radio waves oscillate—only with much more rapid frequencies. Where oscillation is frequent—because of the shape of the basin in which the tide is moving—there may be additional heights.

One of the most famous tides is in the Bay of Fundy between Nova Scotia and Maine off the east coast of the United States and Canada. Tides in the Bay of Fundy may cause variations in the sea of as much as seventy feet. The tide in the Bay of Fundy is called a *resonance* tide, because it is caused by a series of resonances, or back-and-forth movements.

Although all this seems confusing enough, in reality the factors affecting the tides in the various ocean basins are even more complex than this simplified explanation would indicate. The shape of ocean basins, the islands within them, the depth of the ocean, the rotation of the earth, and the passage of the moon and the sun all have to be taken into account in predicting the rise and fall of the tide in any specific part of the earth. Nevertheless, it is possible to issue very accurate tide tables, which tell the height or depth to which the ocean will rise and fall at various points along the shore. Such tide tables are computed in the United States by the U.S. Coast and Geodetic Survey and are made available to the public. They are important in coastal navigation, to fishermen, and for other purposes.

Tides affect the upper levels of the ocean, but ocean water also is affected by the vertical movement of the sea. Upwelling of cold deep seawaters takes place in the open ocean for reasons not completely understood, but also is of importance along the shores of the continents. Probably the upwelling of waters close to shore happens as slow deep ocean currents are pushed upward by the slope of continental shelves.

For example, during the summer months along the Pacific Coast of North America, cold water rises to the surface and meets the warm air lying above it. Often this condenses the moisture in the air, creating a bank of low-lying fog, which hovers offshore during the day and then moves inland as dusk comes. After dawn, the flow of air is reversed, and the bank of fog is pushed back out to sea. San Franciscans call this natural air conditioning, for it keeps San Francisco and the cities around San Francisco Bay as much as 10 degrees cooler than the hot Central Valley of California. The system is so regular that only infrequently does the fog fail to appear. When it does, the coastal cities suffer through the same high air temperatures that affect the Central Valley. In the winter months, the fog bank usually disappears, partly because the air over the ocean becomes colder, but also because upwelling subsides.

A similar current of upwelling cold water rises along the coast of South America. Called the Peru Current, it supplies that part of the ocean with a rich population of fish and the sea-birds that live by eating them. The birds, in turn, deposit their manure or guano on the islands offshore from Peru, islands from which supplies of fertilizer are gathered.

The current is fairly constant, but now and then in summer months it disappears, to be replaced by warmer waters, poor in fish. South Americans call this warmer current El Niño (the Child or the Christ Child), and view its arrival with concern, for it often means economic disaster to fishermen and the guano industry.

El Niño's arrival is unpredictable, although it has happened in 1891, 1925, 1930, 1941, 1951, and 1957. El Niño's cause also is uncertain, but scientists believe it may result from changes in the movement of winds over the sea, which prevent up-welling from the ocean depths.

The appearance and disappearance of El Niño is an example of the close and complex relationship between the atmosphere and the ocean. It is the movement of air which is responsible for much of the movement of the ocean's waters, especially waves. Waves are created by the pressure of the air on the ocean's waters. Although they appear to be movements only of water, in reality ocean waves usually are born far from the shore, created by winds transmitting their energy to water.

As the air presses against the surface of the sea, it pushes drops of water upward. The water rises until it crests, then its droplets fall forward. They then move in a circular pattern, returning to where they had first been affected by the wind. As they fall, they transmit their energy to other droplets of water, which repeat the pattern until it loses its energy or reaches shore.

It is fortunate that water has the ability to transmit the wind's energy in this way and that waves crest and then create troughs. If they did not, waves would reach immense heights. The wind also has a limiting effect on waves after they are formed, topping them off. Although wind-driven waves may reach heights sufficient to break over ships at sea or even lighthouses on occasion, waves crashing on shore usually do not exceed twenty feet.

To the eye of an observer on shore, however, waves appear to be continuous shoreward movements of water. Moreover, waves take complex patterns which often are dependent on the shape of the shore itself. They also have a great effect on the shore, wearing it down, turning cliffs into sand. Like rivers, they continually erode and change the earth's land surface.

In this sense, simple ocean waves are both destructive and constructive. Other purely destructive ocean waves, however, are not caused by winds, but by earthquakes. Sometimes incorrectly called tidal waves, they are tsunamis (from a Japanese word meaning the movement of water in bays or inlets).

Tsunamis have no relationship with the tides. They happen when the bottom of the sea is either raised or lowered by a shift along a deep-sea fault. As the earthquake causes the sea floor to rise or fall, the column of water standing above the fault is disturbed. Should the fault drop, a hole is created in the ocean's surface. Should the fault rise, the column of water above it is pushed upward. In either case, a wave radiates in all directions from the focus of the earthquake, much as the surface of a pond is disturbed when a rock is dropped into it.

Moving at about the speed of a jet aircraft, the tsunami sweeps toward shore. At sea the difference in the level of the ocean is of little importance and usually goes undetected by ships. Near shore, however, the wave runs up the floor of the ocean, raising the level of water along the shoreline by as much as a hundred feet. Often, however, tsunamis are first evident by what seems to be a receding of water from the shoreline, followed by the appearance of a wave which runs up the beach far above the normal high-water mark.

Fortunately, because earthquakes are detected by seismographs and because the speed of tsunamis is now known, it is possible to predict the arrival of most tsunamis before they reach shore. An international seismic sea-wave warning network has been established in the Pacific, the location of most undersea earthquakes, which now issues warnings of all tsunamis to nations around the Pacific Basin. Many hundreds of lives have been saved by the network in the years since it was put into operation.

Like the warning networks in force to chart and follow hurricanes, man has been able to gain a small measure of control over the forces of the sea. But it remains a vast and mysterious part of the biosphere about which much remains to be learned, and it is only in the years since World War II that any large efforts have been made to probe beneath the vacant face of the ocean.

Yet the sea remains reluctantly accessible to man's exploration. Living as we do on land, often far from the shore, we tend to forget that it covers most of the earth's surface, that the sea is as deep as, or deeper than, most of the atmosphere, and that the great engine of the biosphere could not run without the complex interaction of sun, sea, and air. Not only was the ocean the birthplace of all life, it remains the habitat of most of its living organisms. In its depths and pastures lie food and mineral wealth beyond our present comprehension. Through its currents and countercurrents, much of the sun's heat is transmitted to the rest of the world. Without the ability of the ocean to create winds and add water to the air, life on land would be impossible. Just as life began in the sea, so will it continue there, but only so long as mankind respects his power and its ability to nourish the world rather than destroy it.

—5—

CYCLES

Draw a line on a piece of paper. It is easy to see it has a beginning, a middle, and an end. That, in fact, is how you had to draw it, by starting at one place on the paper and progressing across it toward another and ending it there. A line connects two points in the same plane. A line is finite—it *does* have a beginning, a middle, and an end. Much in nature, however, does not. Instead it proceeds in a series of endless cycles, progression through a series of changes, which return to their starting point only to be repeated again.

Indeed, there is some reason to believe that even a straight line, if it could be extended far enough, might eventually return to its beginning. This possibility was suggested by Albert Einstein, one of the most famous physicists of the twentieth century. Einstein proposed that the universe may be a curved continuum so arranged that it has no beginning and no end.

In such a realm, Einstein said, if a straight line were extended far enough, it would reach completely around the continuum from its beginning to begin again. (Einstein's view of the universe was more complicated than this, but for the purposes of the discussion of cycles, this view of the curved universe is sufficient.) Others have suggested that, if one had a telescope of sufficient strength and looked through it, he would see the back of his own neck, another way of viewing this possibility.

It is unlikely the telescope capable of spanning the entire breadth of the universe will ever be invented, but perhaps this is not so important as the idea that the universe is not finite, even for lines drawn within it. Instead, it seems to be a series of interrelated cycles, often interlocking, each of which is constantly repeated. One of the most familiar of these is the cycle of the seasons. As the earth revolves around the sun, its seasons constantly change, but also constantly repeat themselves. In the winter the earth is cold, silent, and without many visible signs of life. With spring, life begins to return again, and the earth warms. Life is at its fullest in the summer and early fall, but gradually falls into dormancy again with the renewed coming of winter.

The cycle of the seasons is dependent on the cycle of the earth around the sun and on the daily cycle of the earth's turning on its own axis. No real beginning or end to either of these cycles can be located, although we arbitrarily declare a year to have begun or ended on a specific date in winter.

Life itself is a cyclical series of events. Organisms are born, pass through infancy and childhood into the reproductive years, and then to death. But while death ends the life of one organism, life itself is constantly repeated. Life cycles vary in length, but the phases remain the same, and all living organisms progress through them. Indeed, were it not for the cyclical nature of life, there would be no living organisms. For it is inherent in life that some members of a species must die to make room for those who will succeed them.

Great variations exist in the length of life cycles. Man's, as an average, may be the biblical three score years and ten, but insects may live only a few weeks, and cells may survive for only a few minutes. Few species live much beyond their reproductive phase, and most species produce far more organisms than will survive into adulthood, probably to ensure that the species will survive.

Reproduction is a key phase in any species. Without it, the species will not continue. Cycles of life, however, are not the only cycles in the earth's biosphere. At least a dozen other major and minor cycles of elements must continue if the biosphere is to maintain life.

One of the most important of these is water. Water has already been considered as a major part of the biosphere, but it is the way in which water is constantly being recycled through the biosphere which adds to its importance.

Water is a unique substance. It may appear in the biosphere as a liquid, as it does in oceans, rivers, and lakes. It may be a solid, as it is in ice and snow. Or it may be a gas—water vapor. Water also is absorbed and fixed—that is, temporarily stored in the bodies of animals and in plants.

Water is an excellent conductor of heat. It also absorbs large amounts of heat without changing its chemical nature or breaking down into its two elements, hydrogen and oxygen. When large amounts of heat are added to it, it cools slowly, releasing its heat gradually.

When it is evaporated, it still retains its molecular composition, but is easily transported through the air. It also is easily condensed again into water, ice, or snow by cooling. Water's cycle can be said to originate in the oceans, and it is to the oceans that it is eventually returned, but before this happens, it often is involved in other cycles. For example, it is an integral part of plant growth. Fixed in plants, it is consumed by animals, where it may participate in their cycles of digestion, respiration, or perspiration.

Water also aids erosion of the land surfaces of the biosphere. As it does, it is involved in the cycle by which minerals are washed from land into the seas. When turned to ice, it also helps to erode the land. Ice forms in tiny cracks in rocks and expands, breaking the rock as it does so. Then as it melts, it moves the broken bits of rock into rivers and eventually grinds them to such fine particles they can be carried to sea as silt and mud.

Over many millions of years, the sediments of rivers deposited along the shores of seas are compressed by succeeding layers of erosion to form sedimentary beds of rock. These, in turn, are driven deeper into the earth's crust and may eventually again become land, which reenters the erosion cycle.

The cycling of water may vary considerably from place to place over the earth's surface. Evaporation of ocean waters and the conversion of water to water vapor is greatest near the equator, where the earth receives more heat than at the poles. It grows progressively less as one moves away from the equator. Even though the polar regions are covered with ice and snow, they actually receive less precipitation than the equator because less water is evaporated from polar seas. In reality, the polar regions are cold, dry deserts covered with a thin layer of frozen water.

The difference in evaporation of ocean waters into the atmosphere makes the time involved in cycling of water vary considerably. As an average, water vapor remains in the air for about nine days, but this is only an average. In some parts of the world, it may remain suspended in the air for much longer periods of time, and in others, notably the tropics, water vapor may be precipitated again to the oceans or the land surfaces of the earth in a shorter time.

Water vapor's passage over the earth also may vary from only a few miles to many thousands. In some parts of the world, water is recycled back and forth locally between the ocean and the air, the air and the land very rapidly. In others, it may be fixed in vegetation for long periods of time, before it finally makes its way back to the ocean.

This frequent recycling occurs because of the fixed amount of water in the biosphere. Were this not true, the earth would have to have a large new supply of water for its purposes from some source constantly. While new water is made, largely because of the interaction of the sun and the atmosphere, most of the earth's supply is used and reused again. This implies that there is a balance of water in the world, that even though water is held in a variety of different forms, each of these forms balances with others.

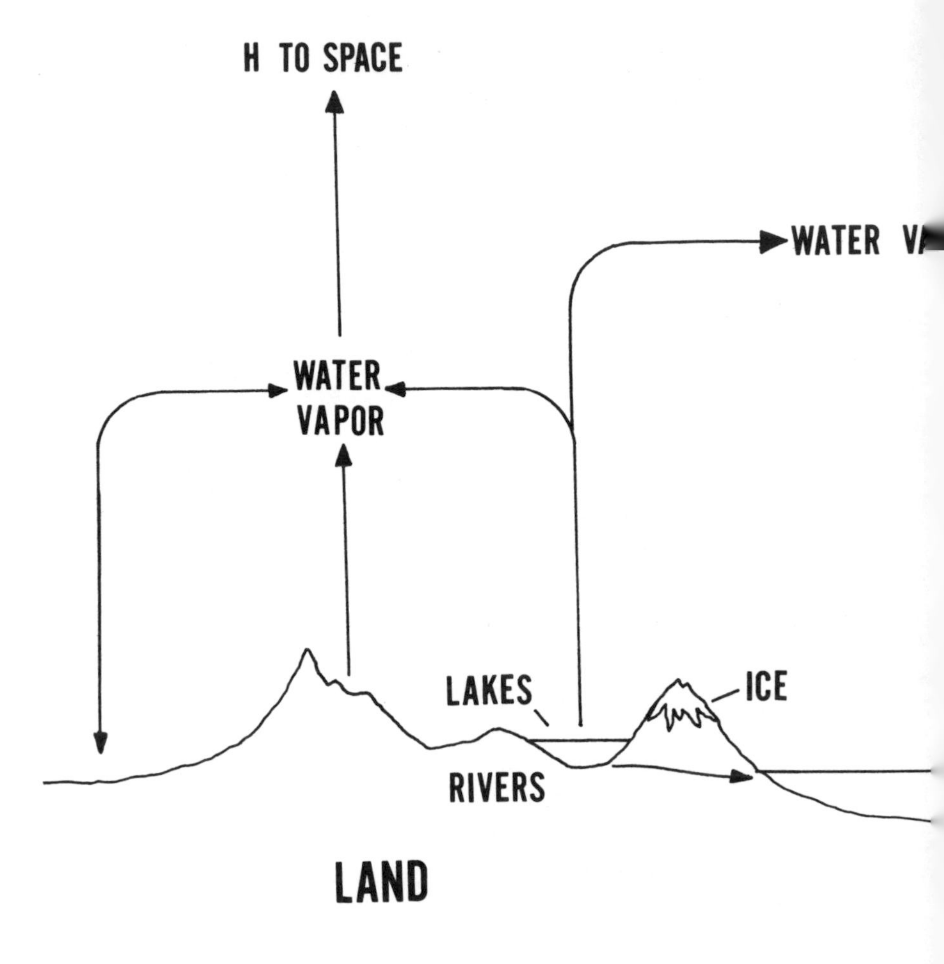

The general circulation of water through the atmosphere, one of the many cycles of elements found in the biosphere. Some hydrogen from water is lost to space, but it is a relatively small amount of the total water circulating in the seas, rivers, lakes, polar ice caps, and as water vapor in the air.

Precisely how much water is held in various parts of the cycle is not certain. Obviously, most of it is in the oceans. A smaller part lies in lakes and rivers, an amount probably about equal to that in the polar ice caps and in glaciers. The smallest part of the water on earth is water vapor, but even though small, this amount is very important. Although it is perhaps only 3 percent of all the water on earth, this water is necessary

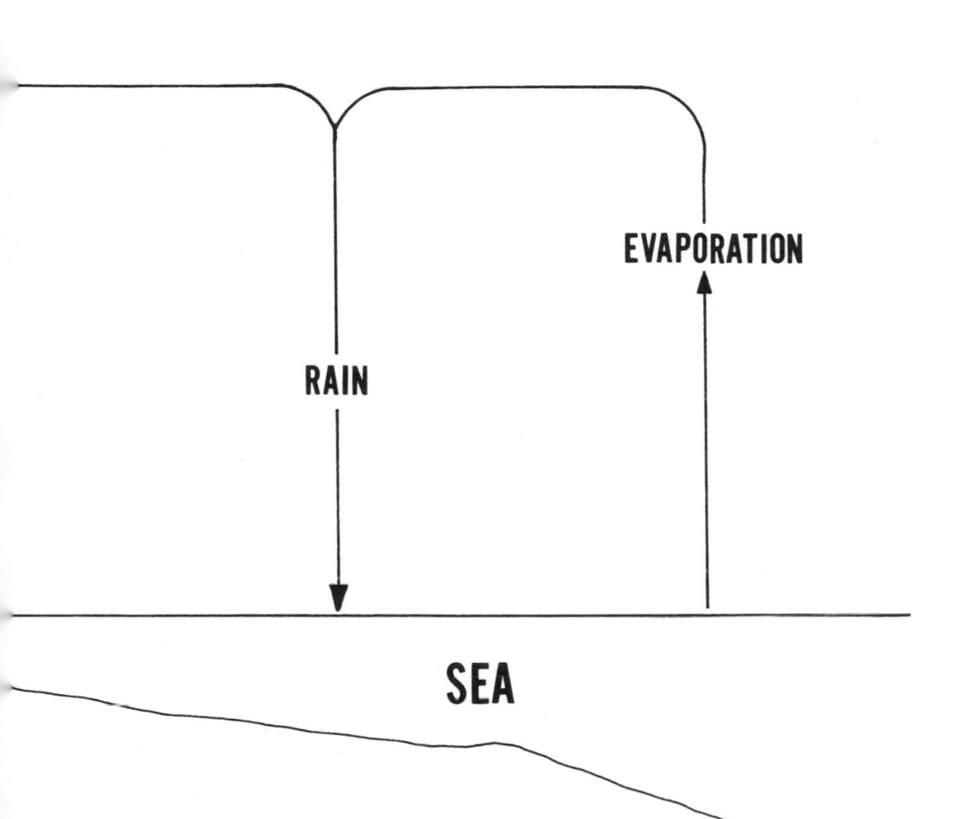

for life on the continents. From it come the rain, snow, and ice needed to sustain nonocean life. Further, only this part of the total water in the world is highly transportable and easily moved from one place to another.

Within the general cycle of water through the biosphere, photosynthesis, the process by which green plants capture and store the sun's energy, is the most important related, interconnected cycle. Photosynthesis is dependent on two other cycles, the transpiration and binding of water. Transpiration is the circulation of water from the air through the soil and up plant roots to be exhausted into the air. Binding is the fixing or storage of water to other molecules in plants and animals. Binding usually is accomplished by the attachment or attraction of the hydrogen atom in the water molecule to other molecules or atoms. Organic molecules—that is, those involved in living plants or animals—are frequently bound by sharing electrons. Often the simple addition or subtraction of atoms or molecules may alter both the appearance and function of a biochemical substance.

All plants absorb water from the soil through their roots, but not all plants absorb water at the same rate, and there is a great difference in the amount of water made available to plants from different kinds of soil.

Men first discovered this fact thousands of years ago, when agriculture came to replace the hunting cultures of prehistoric men. As early civilizations first harvested wild plants and then learned to plant and cultivate them, they discovered the natural groundwater available in the soil often was not sufficient to make plants grow. If early farmers depended only on the available water in the earth, their crops would wilt and die before they could be harvested.

The remedy to this situation was obvious: supply crops with additional amounts of water. Farmers learned to draw water from wells, rivers, and lakes and to carry it to their crops through canals. Irrigation, as this process has come to be called, is now an important part of agriculture in many parts of the world.

Irrigation has proved to be more complicated than simply bringing water to plants. Because of the variety of soils available for growing crops and the difference in their ability to provide water to plants, farmers must know which kinds of plants are best suited to different soils. Some crops grow best in soils which carry little water to their roots. Others require almost continuous irrigation for successful growth. Irrigation, in turn, is often dependent on the natural precipitation of water from the air. The most primitive kind of irrigation is that in which groundwater is supplied completely by nature, as in the Nile Valley, where the annual flooding of the Nile River for centuries provided the necessary water for the crops along its banks. With man's increased ability to use machines, it has been possible to provide irrigation in parts of the world that were formerly desert. Arizona and parts of southern California, for example, now grow crops in what were formerly parched and arid lands because of extensive canal systems built to carry water from faraway sources to the fields.

Ultimately, however, all plant growth is limited by the total amount of water available for transpiration in the biosphere and the natural precipitation of water from the air to the land surfaces of the earth. It is the water balance in the world, then, that determines to a great degree the amount of plant food that man can grow for himself in the biosphere. Natural precipitation is unevenly distributed over the earth's surface. It is greatest near the equator and least near the poles, but other local climates make some parts of even equatorial regions of the earth either wet or dry. The center of northern Africa, the Sahara, is a well-known desert, even though North Africa is surrounded by sea. The center of central Africa, however, because of different climatic conditions, is very wet and the home of tropical rain forest.

India and Southeast Asia are dependent on the seasonal rains called monsoons for the successs of crops, even though irrigation is an important part of farming. In some parts of the world, in spite of heavy rainfall, the evaporation rate of the air is high because of the sun's heat. There, even though the earth receives a great deal of water, it is rapidly taken up again into the air by being heated and turned into water vapor.

Plants themselves add to this problem. Even if water reaches the soil, plants may take it up so rapidly through their roots that they quickly deplete the earth of its moisture. The success of irrigation may depend on the ability of farmers to judge just how much water has to be added to keep the transpiration cycle in continuous operation, and many small applications of water over long periods of time may be more important than a single large dumping of irrigation waters on fields.

Irrigation engineers must distinguish between the two kinds of water available to plants. To do so they call the water which is in the earth—but not always available to plants—groundwater. That which can be taken up by plant roots, probably a smaller quantity of the total water in the earth, is called soil water. Groundwaters do little good unless they become available to plants as soil water, except as a measure of how wet the landscape in general may be and how soon rivers or lakes may become useless as sources of irrigation water.

Soil water obviously depends on how deep plant roots can reach to obtain the water they need to stay alive. The kind of soil thus becomes an important part of deciding which crops to plant and how much they must be irrigated. In some parts of the world, it is probably impossible to grow more plants per acre (and thus gain more food). Even though water is available, it cannot be gotten to plant roots. In others, the inefficiency with which water is provided to crops is so great that even if more water is used in irrigation, it is not possible to grow more food. This is particularly true in Asia and Africa, where primitive farming methods are still used.

Crop production also is involved in the cycling of nitrogen through the biosphere. Nitrogen is an element needed by plants to make protein. When the plants are eaten by animals, including man, the nitrogen may be used to make new kinds of protein required for life. Nitrogen is abundant as a gas in the atmosphere—indeed, 79 percent of the air is this element. In the atmosphere, however, nitrogen cannot as a rule be taken up by plants or animals and used to make protein. Legumes, such as peas, beans, peanuts, and clover, for instance, are the only plants that can take nitrogen from the soil. For most plants, nitrogen must be fixed or stored in the earth.

Until the beginning of the present century, nitrogen was almost entirely fixed by plants or by certain microorganisms, but during World War I and in the years following, various processes were developed to fix nitrogen artifically. Man began to produce large amounts of nitrogen fertilizers and to feed these into the ground in efforts to increase food production.

Vast amounts of nitrogen have been fixed in this way since the 1940's, so much so that the amount has increased sixfold each year for the past decade or so. Some experts now predict that by the end of the twentieth century, more than 100 million tons of nitrogen will be added to the soil every single year.

Just what this is doing to the balance of nitrogen in the biosphere and to the nitrogen cycle is not clear, because scientists do not know how long it takes to cycle the element through the biosphere under natural conditions or the amounts of nitrogen fixed at the various stages of the cycle.

Nitrogen's cycle is a complicated one. Some new nitrogen is released into the atmosphere each year by volcanoes, but it is an amount at present impossible to measure. An even smaller amount may come from the action of cosmic rays and lightning on the upper levels of the atmosphere, but most natural nitrogen is made available for living organisms because of its passage through microscopic bacteria and algae. Some algae are able to live directly on nitrogen, but most natural nitrogen becomes available from the action of bacteria in decay processes.

Eventually, it is believed, the soil releases nitrogen to the air, but how and in what quantities is not known. Because of this, it is impossible to tell how harmful man's tampering with the natural cycling of nitrogen in the biosphere may be. The only known examples of too much nitrogen in the biosphere are to be found in lakes or streams where sudden large amounts of it have produced "blooms"—sudden large growths—of blue-green algae. Blooms of nitrogen often so increase the amount of algae in freshwaters that all other life disappears. But whether too much nitrogen in other parts of the biosphere would have the same effect remains to be seen.

An additional complication of nitrogen cycling is the large amount of heat energy needed to extract nitrogen from the air to make nitrogen fertilizers. Although the use of nitrogen fertilizers increases the growth of crops, often the cost in energy for such growth is almost as much as the increased production yields. Nitrogen may also be introduced into the soil by planting species of plants called legumes. Legumes not only produce food, they also fix nitrogen for use by other crops. Unfortunately, this, too, requires an expenditure of energy at a cost nearly equal to that realized from the additional crops produced.

Thus man faces two uncertainties in the nitrogen cycle: He does not know if he is removing too much of this vital element from the atmosphere, and he is uncertain as to whether such action is polluting the biosphere to dangerous levels.

A second key element being cycled through the biosphere is oxygen. Oxygen makes up 21 percent of the air. It is basic to all living matter and is required for certain basic biochemical reactions without which life could not continue. Indeed, one quarter of all living matter is believed to be made of oxygen atoms.

Yet oxygen may not always have been so abundant in the atmosphere. Only 50 million years ago—a relatively short time ago in the earth's long history—the air probably contained considerably less oxygen than it does today, and early in the earth's life, the air contained almost no free oxygen at all. Most of it was imprisoned in rocks. The oxygen then in the atmosphere was combined with carbon to make carbon dioxide.

Since then the amount of free oxygen in the air has gradually increased because of biologic activity. As plants evolved on the surface of the continents, more and more oxygen was freed by their reduction of carbon dioxide into carbon and oxygen, setting up the photosynthetic process now in balance in the biosphere.

Oxygen also was freed by the breaking down of rocks on the continental surfaces, a process which continues today. Volcanoes also release some oxygen into the atmosphere, although both of these sources of free oxygen probably are less than that provided by plants. Plankton in the sea also release oxygen and absorb carbon dioxide, and ultraviolet light helps to free additional amounts from water vapor in the atmosphere.

The way in which each of these subcycles of the oxygen cycle works, like that of the nitrogen cycle, is a complicated set of interlocking relationships about which very little is known. However they work, they have combined to increase the amount of free oxygen in the atmosphere during the last 100 to 50 million years. The increasing numbers of humans on earth have had a paradoxic effect on free oxygen. Man has become one of the great consumers of oxygen. At the same time, he has helped to increase its levels by planting crops. His consumption of oxygen is a part of his normal breathing process, but man also has worked to reduce the amount of free oxygen in the air by burning vast amounts of fossil fuels. By burning carbon-rich fuels, man has added large amounts of carbon dioxide to the air.

The ultimate effect of this process is as uncertain as the ways in which oxygen is released into the atmosphere, but certainly too much carbon dioxide may have serious effects on the earth. Too much carbon dioxide could help to heat the air. It might cause the polar ice caps to melt, raising the levels of the sea, or it could prevent sunlight from reaching the earth's surface—as seems to happen on Venus—creating a kind of greenhouse in the atmosphere.

It also is not clear what the results of the freeing of even more oxygen from the earth and placing it in the atmosphere would have on the earth. Too much oxygen can poison plant life and can kill humans, but not enough is known about oxygen's effect on other organisms to be sure that this would cause a general decline in animals. Such a possibility does not seem likely. Most of the oxygen on earth is in its rocks. Smaller amounts are held in the seas in combination with other elements. The balance of all the oxygen in the world is as uncertain as are the balances for other cycling elements in the biosphere.

The oxygen cycle also moves very slowly, and some scientists have estimated that it may take as much as 2 million years for oxygen atoms to pass from the atmosphere into rocks and then to be freed again as oxygen in the air.

The cycling of carbon is mixed with that of oxygen because an important pathway for carbon's passage through the biosphere is as carbon dioxide. Carbon dioxide is involved in the cycling of carbon on land. But carbon also is cycled in a different way in the oceans. There it is assimilated by plankton, which draw their nutrition from ocean waters. Minute single-celled animals, they are eaten by larger creatures in the sea. When plankton die they sink to the ocean bottom, where they eventually are compressed into rock or become parts of beds of oil and coal. Other dead plankton are lifted to the sea's surface by the upwelling of deeper waters. As decayed materials, they eventually break down into nutrient substances, which can become a part of the cycle again after being absorbed by new generations of plankton near the sea's surface.

On land, carbon is exhausted into the air by plants, notably by the world's forests, which give off carbon dioxide through their leaves. Lesser amounts of carbon dioxide are put in the atmosphere by animals. The soil also releases carbon dioxide. Plants absorb carbon dioxide, principally during the day when they are engaged in photosynthesis, but they also store carbon, both in themselves and in the soil around them. Plants eaten by animals are a part of the carbon cycle, too. They make it possible for carbon to be moved from the land into animals. Both dead plants and animals release carbon into the soil as a part of decay, completing the general cycle of carbon outside the oceans, and also contributing to the creation of pools of fossil fuels underground.

Since the middle of the nineteenth century, man entered this cycle by burning large amounts of fossil fuels, thus putting carbon and carbon dioxide into the atmosphere. He has also cut down a large part of the earth's primordial forests. This has removed a place where carbon can be stored—trees—and turned this carbon into the atmosphere, where it has been added to the carbon dioxide already there.

As with the other natural cycles of the biosphere, it is impossible to predict the effect of this change or to say whether it is profound or insignificant. Few of man's tinkerings with cycles appear to be of little consequence, however, and it is likely nature has extracted a price for this change, just as it has for others. We must wait to see how high that price may be, not only for carbon's use, but for our interference with others as well. Our survival may depend on how quickly we learn that price and how willing or able we are to repay it.

—6—

ICE

If you have ever been to Yosemite National Park in California, you cannot help but be impressed by its cliffs and magnificent rock walls. Yosemite Valley, more than three thousand feet deep, is a great furrow in the granite underlying the Sierra Nevada Mountains. Its sheer edges seem to have been cut by giant knives.

When you stand at the bottom of the valley, you may find it difficult to believe that Yosemite was carved slowly and patiently by ice. Once a glacier filled the valley to its rim. Then, thousands of years ago, the earth warmed, and the glacier slowly retreated up the valley and disappeared. Today all that remains of it is a tiny fragment on the slopes of Mount Lyell, some distance from the valley itself.

Ice and glaciers have left their imprint on many parts of North America and northern Europe. Although a few glaciers remain, most have melted away, leaving their paths in the rock as footprints of their passing.

Glaciers are reminders of a time when the world was once much colder than it is today. They represent one of the unsolved mysteries of geology. Much is known about how glaciers form, flow, and retreat, but little about why the ice age that they represent visited the earth. Glaciers last flourished in the Northern Hemisphere more than ten thousand years ago and probably reached their height more than fifteen thousand years ago.

The difference between the world then and now is almost beyond our comprehension. During the height of the ice age, great domes of ice lay over both North America and Europe, covering more than 9 percent of the earth's land surface. Temperatures averaged 15 degrees Fahrenheit lower than they do now. Central Europe also was covered with ice, and ice caps reached as far south as 38 degrees north latitude in North America, far into what is now the temperate zone of the world.

The extension of the ice also had a great effect on the rest of the world. Rain fell constantly in the Sahara, a desert where now there is often no rainfall for years. The increase in the volume of ice also had a great effect on the oceans. As more and more of their water was frozen, their level dropped as much as three hundred feet.

Then the ice began to disappear as mysteriously as it had appeared. Although a miniature ice age may have developed about five thousand years ago, most of the world grew gradually warmer and warmer until the earth of today came into being. Like the ice age, today's present warm interval is not permanent, however. A new gradual decline in temperature, whether temporary or long-lasting, now seems in effect, and the world is growing colder.

The last ice age, moreover, was not the only one to have been visited on the earth. At least three or four times in the past, ice has covered large parts of the globe. On each occasion, the ice age was followed by an interval of warm weather. During some of these warm intervals, the earth was warmer than it is today and had even less ice than now is frozen in the polar ice caps.

The extension of ice into the temperate zones of the world has also had profound effects on life in the biosphere. At the height of the last ice age, the northern latitudes of the earth were so inhospitable that most life was forced into a narrow band around the equator. The cold also seems to have made whole species disappear into extinction, notably the dinosaurs. These giant reptilian creatures were cold-blooded, and probably unable to survive in a cold climate. They gave way to mammals, warm-blooded animals which could maintain a constant interior body temperature and so live through the long winter the world experienced.

The drop in the ocean levels also had its effect on life. It created a series of land bridges between islands and continents over which living creatures could migrate in search of warmer homes. One of the most important of these was probably what is now the floor of the Bering Sea between Alaska and Siberia. Whether land or ice, the Bering Strait bridge made it possible for man to move eastward out of Asia and Africa, his ancestral homes, into North and South America. Exidence shows such a migration. Primates do not seem to have been native to either North or South America. Instead they seem to have come late to these two continents over a land or ice bridge.

As the sea level dropped, it may also have helped to form peculiar flat-topped undersea peaks called guyots. These strange undersea features dot the floor of the Pacific Ocean. Geologists believe guyots were formed when seamounts, undersea mountains, were thrust above the ocean as its level dropped, eroded by wind and water, and then covered again as the ice melted and the oceans rose.

How did so much ice come to form? The answer is not known. The creation of ice and of glaciers is understood, however. Ice forms when snow from a preceding winter fails to melt in the summer. As new snow falls, the lower layers of the snow are packed ever more densely until they become ice. Some ice forms in the highest mountains on earth today in this way, but the world's climate is not sufficiently cold at present to make this a general pattern. The coming of the ice ages then was the result of much colder general weather in the world, weather cold enough so that snow was not able to melt for many years. When sufficient ice had been formed, it became a glacier.

Glaciers arise in mountains. As they form, they obey the law of gravity and seek a lower elevation, just as water in liquid form does. Slowly the glacier begins to move downhill, carving away rock, soil, almost anything in its path. As long as a glacier is fed by new snowfalls at its source, it will continue to follow a downhill path. Only when the weather warms does the ice halt and then begin to retreat, leaving behind the debris it has ground beneath it—a moraine—as proof of its passage.

Why did great amounts of ice form into many glaciers and into very thick and extensive ice caps? Many theories have been advanced, but none of them is completely satisfactory. Such variations in climate were first brought to the attention of geologists early in this century by Swedish oceanographer Otto Pettersson.

Pettersson lived on the shore of the Baltic Sea. There he became interested in peculiar undersea tides, which he was able to measure along the coast of his homeland. The tides arrived at twelve-hour intervals. Pettersson's tides are not to be confused with the conventional tides, which raise and lower the level of the ocean. His tides seem to be swells of cold water that arrive at the shore by passing under layers of warmer water.

Pettersson concluded the tides were somehow caused by movement of undersea waters against or over the Mid-Atlantic Ridge, a range of undersea mountains which run north and south down the middle of the Atlantic Ocean. He believed they were also the result of movement of the earth, the sun, and the moon. At least they seemed to follow varying relationships among these three bodies.

The variations in temperature also seemed to have an effect on the arrival and disappearance of herring, a fish important to the Swedish fishing industry. In an effort to find out why such cold waters seemed to ebb and flow, Pettersson investigated past history for records of changes in climate over the North Atlantic. Somewhat to his surprise, he found historical records showed several important, gradual cycles in the general weather in this part of the world.

For example, during the time when Viking invaders actively

sailed the North Atlantic, the weather was much warmer than it is today. Pettersson was able to determine that during this time, drift ice—ice that has broken off the polar ice cap and drifted south into the North Atlantic—was nonexistent. Its absence permitted Norsemen to cross the North Atlantic to Greenland, Iceland, and probably North America. Then as mysteriously as it had disappeared, the drift ice reappeared, and the earth grew gradually colder. The cold reached a peak, and then the world began to warm again, only to be followed by a second local cooling. This was followed by a warming trend that reached its height in the 1940's. Colder weather now seems to be coming again.

Pettersson was not sure why these variations in temperature took place, but he published his findings in a paper entitled "Climatic Variations in Historic and Prehistoric Time." The paper helped to bring the attention of other scientists to the problem of great general variations in the earth's climate.

Since then, a number of suggestions have been made to explain the coming and going of the ice and its effect on life in the biosphere.

One of the most interesting is the effect of sun spots on the earth's climate. Sun spots are cool islands that appear and disappear on the sun's surface from time to time. They seem to follow a regular cycle of eleven years, interrupted by periods of two years when the sun is quiet and no spots appear. Their cause is not known, but John A. Eddy of the U.S. High Altitude Observatory in Boulder, Colorado, believes they may be the result of interaction between the sun's magnetic field and convection currents inside the sun. Convection currents bring the sun's hot gases from its center to its surface. There they may be affected by the magnetic fields that surround the sun as they do the earth.

Dr. Eddy also believes there is enough historical evidence to show that sun spots may disappear for relatively long periods of time. For example, he points out that during the seventeenth and eighteenth centuries, the world was much colder than it is today. From 1645 to 1715, Europe suffered through a cold wave so severe that rivers froze, glaciers advanced, and rain and snow were more general than at present.

Dr. Eddy says that an examination of old maps of the sun's surface prepared between 1642 and 1644 by the Polish astronomer Johannes Hevelius show that during this period, just before the onset of the cold weather, sun spots were conspicuously different. They moved more slowly across the surface of the sun, because the sun was revolving more slowly than it does today. The sun, like the earth, revolves on its own axis, but it does not turn uniformly. The movement of rotation is slower at the poles of the sun than it is at its equator. During the time when Hevelius was charting sun spots, rotation appears to have been slowed by as much as a day.

Although no other such historical solar records are available, the same thing may also have happened between 1460 and 1550. During this time, too, the earth was much colder.

Dr. Eddy says it is unreasonable to believe that because the sun spots now appear in eleven-year cycles that this has always been the case. He also points out that during the years in question, historical records also indicate the northern lights were not seen because there was not enough solar activity to feed energetic particles into the solar wind. The northern lights are dependent on the coming and going of sun spots, and their absence would temporarily halt the appearance of these spectacular lights in the northern skies of the world.

Still another suggestion for changes in the earth's general climate revolves around the eruption of volcanoes. Volcanoes send large amounts of dust and ash into the upper levels of the atmosphere, which may be suspended there for months or years. A funneling of debris into the atmosphere of this kind took place in 1883, when the volcano Krakatoa erupted in the Dutch East Indies. Krakatoa was one of the greatest volcanic eruptions in historical times. It sent a shock wave six times around the earth, and the vast amount of dust it discharged into the air created spectacular sunsets for several years after the eruption. The climate of the earth also seemed affected by the eruption, largely because volcanic dust has the double effect of shielding the surface of the planet from the sun and of reradiating the sunlight that does reach the surface back and forth through the air. This makes the atmosphere somewhat like a greenhouse, trapping heat and making it more likely that water vapor suspended in the air will be precipitated as rain or snow.

Volcanoes once were more active than they are today. Although this was millions of years ago, their eruptions may have had the effect of creating atmospheric conditions suitable for long periods of cold, snowy weather.

A further clue to the coming of the cold lies in the polarity of the earth's rocks. Polarity is fixed in rocks when they cool to a specific temperature, called the Curie point. Such cooling aligns volcanic rock in a north–south direction when it reaches the Curie point. Above the Curie point, volcanic rock will have no polarity. Below it, the rock will be in line with the magnetic alignment of the earth at the time at which this temperature is reached. Examination of the Pacific ocean floor by Dr. Victor Vacquier of the Scripps Institution of Oceanography in 1958 showed in places it is composed of alternating bands of rock with opposing polarity. These bands are at distances from one another, but seem to match. Later investigations have shown that the convection currents of the earth's mantle are thrusting hot material up through places in its crust. As this material reaches the floor of the sea, it spreads in opposing directions, is cooled, and is fixed with the polarity of the earth at that time. Thus, the polarity of the earth has varied in opposing directions at different times in the past.

This must mean that the magnetic poles of the earth also have reversed themselves over many millions of years. Why this has happened is less clear. Its effect on the climate of the world is also uncertain, but a change of polarity might mean that the earth's poles have wandered over the face of the planet in ages past. Some other geologic findings also indicate this to be true. Antarctica is a continent, but most of its surface is masked by a thick polar ice cap as much as a mile deep in places. Where the ice does not cover the continent, however, coal desposits have been found. Coal comes from fossil plants, and plants could not have grown in Antarctica unless it was once much warmer than it is today and free of ice. It could not have been subject to these conditions unless the world was so warm the polar ice caps had all but disappeared or the poles were in places on the earth's surface different from today's.

Some scientists have also suggested that slight variations in the earth's rotation on its own axis, or in the orbit which it travels around the sun, may have affected the climate in the past.

Such a theory has been outlined by Nicholas Shackleton of Cambridge University. Dr. Shackleton has used a computer to chart the variations in the amount of heavy hydrogen found in tiny shells left in sea sediments. Shells are deposited on the ocean bottom after their inhabitants die in upper ocean waters. During times when the sea's waters are cold, the amount of heavy hydrogen in the shells increases, probably because much of the ocean's water is taken up into ice and that which remains liquid contains more hydrogen. Shackleton's studies show a regular rhythm to the coming and going of ice, a far more regular variation than some other scientists have found.

Shackleton believes these variations match small but important changes in the angle at which the earth is tilted on its axis and in the form which the earth's orbit takes around the sun, a theory based on work done in the 1920's and 1930's by a Yugoslav scientist, Milutin Milanovitch.

Although the earth's angle of inclination of its axis from the perpendicular is usually given as about 23.5 degrees, it actually wobbles slightly over many years with an angle that varies from 21.8 to 24.4 degrees. The variation takes place over a 40,000-year period. At the same time, the earth's orbit around the sun shifts from a slight ellipse to almost a perfect circle. This means that while the earth is moving in an ellipse, it may be three million miles closer to the sun at some times of the year than at others. When its orbit is a circle, of course, it remains at the same distance from the sun throughout the year.

The Milanovitch–Shackleton theory holds that the combination of these two events may cause significant differences in the amount of sunlight which the earth has received during the past. When the wobble and the orbit coincide, this may mean as much as 30 percent less sun than at other "normal" times. Shackleton's measurements of seashells matches the variations in the earth's position with relation to the sun, seeming to indicate that the ice ages are a result of the variations, even though slight, of the earth's inclination and distance from our star.

If this is true, it means there is not only an explanation for ice ages, but a method of predicting them. It also seems to indicate that there are three general patterns to the warming and cooling of the earth. The first are the great periodic swings, however they are made, which bring prolonged ice ages. These happen over thousands of years and have immense effects on the surface of the world and on the life within the biosphere. The second are shorter periods of warm and cold, like those Pettersson discovered by examining historical records of the past. These seem to last for only a century or two, and while significant to life, do not alter it so drastically as a full ice age. Finally, there are shorter local changes in the weather, which affect specific parts of the world.

One of the latter took place in the 1960's and 1970's in the Sahel, the southern fringe of the Sahara Desert in Africa. The Sahel experienced a drought of about seven years, beginning late in the 1960's, during which almost no rain fell. There was a widespread failure of grass, and many animals and some humans died of starvation.

Droughts have afflicted the Sahel before—in 1900, 1911, and 1930—but the most recent failure of rainfall was the most serious in this century. Old wells dried up, the water level in new wells dropped, and had it not been for relief efforts from other nations, there might have been more severe loss of human life. The situation in the Sahel during the most recent drought was complicated by other problems, but even so, it is an indication of how local weather conditions can seriously affect parts of the biosphere even when the general climate is stable.

The cause of the Sahel drought is not known. Some scientists believe the lack of rainfall resulted from a general cooling trend, the same trend which began in the 1940's and has been continuing in the Northern Hemisphere ever since. Others think the drought is only a local peculiarity of weather, complicated by the way in which farming life in the area has changed.

Whatever the cause, local variations in rainfall do exist. In the Great Basin, the states between the Rocky Mountains and the Sierra Nevada, periodic variations in rainfall dating back thousands of years have been found by measuring the thickness or thinness of tree rings in bristlecone pines. Bristlecone pines are found only in the White Mountains, an arid range of peaks on the border of California and Nevada. The pines are able to survive on almost no water and are among the oldest living things on earth. Some of those found in the Bristlecone National Forest are thousands of years old. Thus their rings contain a record of how much rain has fallen for centuries. As with the Sahel, the reason for cycles of increasing and decreasing rain is not known, but it is clear the cycles vary over short periods of time, about fifteen years.

To discover the reasons for both local and long-range changes in the climate of the biosphere, it will be necessary to make many more measurements similar to those of seashells and tree rings.

More also needs to be known about how the currents of the ocean affect the climate of the earth. Because they are water and because they absorb so much of the radiant energy which the sun showers on the earth, they must be important both in the formation of general and local climatic conditions. Most of the ocean's absorbed heat, however, appears to be stored in surface waters. Beneath them the water is much colder and may move less rapidly than do surface currents. The difference between warm and cold water has an effect on the circulation of seawaters and on the circulation of the air above them.

Just as warm air rises and cold air sinks, so does warm water rise and cold water sink, eventually reversing its position with the water at the surface and substituting for the colder water at the bottom of the ocean. But this kind of circulation is not anywhere near as rapid as the movement of masses of air, nor is much known about how it is accomplished. It is likely that warm water flows north or south from the equator after being heated by the sun. Then, as it cools, it sinks to the bottom. But it may be carried for hundreds or thousands of miles before it is sufficiently warm to rise again toward the surface of the sea.

The amount of time such "turning over" takes is still a matter of study by oceanographers, but by measuring the radioactive carbon content of water, just as it is used on land to measure age, it appears that about once every thousand years, surface waters make a complete cycle. Many measurements of deep seawaters, and the rate at which they move, must be made before it will be possible to be more precise about deep-sea circulation.

Obviously, if the amount of heat absorbed by the ocean declined, even by a small amount, it would bring a corresponding drop in the average temperature of the ocean. Some scientists have calculated a decrease of as little as 8 percent of the sun's daily output would cause the average temperature to fall to 41 degrees, the average temperature during the coldest part of ice ages. A decline in the general temperature of the air then would bring another drop in temperature on earth and another ice age.

Yet another theory for the evolution of ice ages is based on the recent information obtained from the drilling of the deep sea floor by the oceanographic ship the *Glomar Challenger*. Cores of the sea bottom and measurements of sea-floor sediments within them seem to show that the coming of ice to the world may have resulted from the movement of the continents over the earth's surface. Most geologists now agree that either the continents were once much closer to one another or they may have been a part of a single continent which broke apart.

When the continents were closer together or were a single land mass, the circulation of the ocean currents was much more uniform than it is today. They may have been so uniform that the heat showered on the earth's equator by the sun was much more evenly distributed among the ocean's waters. Then currents could flow unimpeded north and south from the center of the earth toward the poles, warming them and keeping the amount of ice there at a minimum.

As the continents drifted apart, however, this no longer was true. North and South America gradually drifted westward from Europe and Africa, creating the Atlantic Ocean. This set up a different flow to the ocean's waters. In the Atlantic, it allowed the cold waters of the Arctic Ocean to reach the northern shores of Europe and North America, and it made the Circumarctic Current the only truly global current. Its waters, however, were not able to mix well with the warmer waters of the Atlantic. Instead, they circulated cold water around the continent of Antarctica, making possible the gradual accumulation of more and more snow and ice, creating the ice cap we know today.

The continents are still shifting around on the earth's surface and will take up new positions in geologic eons ahead. Presumably this, too, may alter the flow of the oceans and the circulation of heat through the sea, but only over millions of years. The effect of such continental movement is difficult to predict, just as it is difficult to say when ice will return to the earth again in larger amounts. A more likely reason for the formation of new ice, however, may be not continental drift but the continued addition of heat to the atmosphere. Although it is not possible to say what might cause such an increase, its effect would be another ice age. This is because there is a limit to the amount of heat the air will absorb—just as there is a limit to the amount of heat water will absorb, its boiling point. When air gets too hot, it begins to condense the water vapor it contains into rain or snow. If the air were heated too much over large parts of the earth, it could greatly increase the amount of snow dropped on the land surfaces of the world. As more and more snow fell, ice would form, and a new ice age would be on the way. Because most of the heat the earth receives comes from the sun, it is logical to assume that the beginning of a new ice age would result from a sudden increase in solar energy. But this may not be true. It is also possible that man himself may overheat the atmosphere.

Greatly increased burning by man during the past half century, an amount far greater than has been used on the earth's surface since it was formed, has dumped large amounts of heat into the atmosphere and, to a lesser degree, into the rivers, lakes, and oceans. If the present rate of energy consumption continues, within the next century the average temperature of the air will rise by 2 or 3 degrees. Some scientists believe that if the average air temperature were raised an average of 5 degrees Fahrenheit during the winters of a single century, man himself might heat the air sufficiently to bring a new ice age.

As a matter of record, the average wintertime temperature has been rising recently each year, but it is not possible for scientists to say whether this is because of man's heat contribution to the atmosphere, or whether it is a permanent rise. On the other hand, no scientists would deny that the increasing consumption of energy has released great amounts of heat into the air and that there is much uncertainty as to what effect this burning of fuel has had.

Future calculations of temperature are also uncertain because the numbers of men are increasing at a rate far greater than at any time in the past. The population of the world will soon rise above 4 billion, and this number may double before the end of the present century. As more human beings are born, they will require the burning of more and more heat with a greater and greater addition of heat to the atmosphere. Then the rates at which man is presently adding heat to the air will increase even more.

Thus, for the first time, man as well as nature may have a long-range effect on the world's climate. This only adds uncertainty to an already uncertain future for the long-term changes in the biosphere. If history is any teacher, however, we must believe that the long cold of a renewed ice age will visit the world at some time in the future. The question is not whether this is likely to happen, but how soon it will take place.

—7—
ZONES

Near the tops of the highest mountains live lichens, crust-like combinations of primitive fungi and algae that grow on rock faces, even in the harshest of weather. Lichens are able to stand great extremes of cold, dryness, and heat. They absorb water from the air and extract nourishment from the rocks to which they cling. By secreting acids, they are able to eat into the rock face, both to secure a foothold and to gather minerals for survival. Storms help them to spread by blowing their spores to other rocks. Occasionally, whole pieces may be carried by the wind from one place to another.

Lichens live at the upper limit of the biosphere, just as starfish, worms, and some other simple creatures survive at the bottom of the deepest part of the sea. Between these two extremes are the rest of the plants, mammals, fish, birds, microorganisms, and many other forms of life that populate the biosphere. For the biosphere is filled with life. Indeed, its very reason for being is to nourish and continue life. Were it not for the interplay between plants and animals, the biosphere would cease to function as a place where living things can survive.

Yet despite the presence of life in almost every part of the atmosphere, the sea, and the continents, life is not distributed uniformly over the earth's surface. Nor can much of it live at every level of altitude or at every depth in the sea. Because life is dependent on the driving forces of the biosphere—its winds, waters, the circulation of its gases, and on the heat of the sun—it flourishes only where these factors are most favorable. It is only a generalization to say the biosphere is filled with life. It would be more correct to say life is distributed through the biosphere and that this distribution tends to change as the forces that nourish it change.

As one examines the location of life on earth, two things become clear. The first is that life is located in regions or zones. The second is that like kinds of creatures tend to group together and to be associated with other groups of creatures necessary for survival. One group tends to live upon another. Because this is so, it is possible to look at the biosphere with two different sets of standards. One is based on the kinds of life, the other on where they are located on the earth's surface.

In general, life is most abundant and varied near the earth's equator, because the middle of the earth is generally warmer than other regions, receives more rainfall, and hence is more hospitable to life. As one moves away from the equator toward the poles, life tends to decrease both in numbers and in numbers of species. This is true both of plant and of animal life.

Such a general view of the biosphere, however, does not tell much about specific places on the earth's surface. To understand individual parts of the biosphere better, scientists have sought to relate similar kinds of communities and their place in the biosphere into a unified view of the living world. This study is called ecology.

Ecology is a much-used and much-abused word. Often taken to mean the conservation of natural life, strictly defined it is the science of examining the places where life exists and the different, but often interrelated, kinds of life in such places.

The first great ecological division of the earth's surface is that of realms. Realms are geographic divisions based on the species found in them and not duplicated in other realms. They were first suggested by the English naturalist Alfred Russel Wallace, one of the proposers, with Charles Darwin, of the theory of natural selection. Wallace was able to draw a line dividing parts of Asia and parts of the East Indies (now Indonesia) by charting the location of species within them. The line separated one set of species from another.

Since then, ecologists have defined six realms:

The Nearctic—all of the lands of North America roughly north of Central America;

the Palaearctic—Europe, most of Asia (except parts of the Arabian Peninsula, India, and the Malay Peninsula), and including North Africa north of the Sahara;

the Oriental—India, the Malay Peninsula, western Indonesia, a small part of southern China, and the Pacific islands north of New Guinea;

the Australian—Australia, New Zealand, New Guinea, and adjacent islands, including eastern Indonesia.

the Ethiopian—Africa south of the Sahara and parts of the Arabian Peninsula;

the Neotropical—South America, the West Indies, and Central America.

Though ecological realms are of importance to ecologists, they have little meaning to the average person. What is more understandable to them are the zones within each realm, called biomes. There are eight biomes (although not all realms contain all eight): coral reef, desert, tropical forest, temperate deciduous forest, rocky coast, tundra, grasslands, and the sea.

Although not all realms contain all eight biomes, where biomes exist they are similar in climate, plant and animal life, and geologic conditions. For example, the Mojave Desert of California, Arizona, and Nevada is similar to the Sahara Desert of Africa. The tropical forest of South America is similar to the tropical forest of Africa. Species in different biomes are not precisely duplicates, but they are much the same.

Within biomes are the habitats, the ecological niches where individual species exist. Habitats may be very small—many different kinds of mites, for example, may be found on a single bird—or they may be quite large. One of the world's largest habitats was the nineteenth-century range of the American bison, or buffalo, when it roamed over most of the Great Plains.

Habitats should not be confused with communities—places where combinations of different species live together in the same general area, often dependent upon one another for survival. Communities usually are in a state of constant change. Often one species is arriving as another is disappearing. Change may result from the natural forces of the environment or, more frequently today, from the intervention of man, a creature which transcends all realms, biomes, habitats, and communities.

When change takes place and one species follows another in a community, a succession is said to have occurred. Succession is taking place all the time in the biosphere. It usually is unpredictable. When it causes the disappearance of a species, it is said to have caused extinction. Extinction does not take place frequently, but it, too, is happening all the time.

Hawaii is an example of a place where many successions have happened, especially since the arrival of man. Once the Hawaiian Islands were barren cones of volcanoes that rose above the surface of the sea by a long series of eruptions. They were without life at first, but gradually water and wind brought a few plants and animals. Then men migrating north from Tahiti landed on their shores, bringing with them new species. In the nineteenth and twentieth centuries, more and more men arrived on the islands, adding more and more different kinds of life. One species after another was introduced. Some found themselves without the natural enemies they had had in their previous homes, and they multiplied rapidly, usually at the expense of earlier species.

Succession after succession has followed until some of the first forms of life to reach Hawaii, both plant and animal, are now almost or completely extinct.

Not all successions are destructive to a community, however. In some cases, its various members are able to achieve an equilibrium, a balance, in which the different life forms can exist side by side in a habitat, each with a stable population.

When this happens, a community is said to have achieved a climax state. Nature seems to seek the establishment of a climax community, although it seldom achieves it. Even when it does, the balance between the forces of the biosphere and the various populations in the community often is a fragile one, easily disturbed by the removal or addition of only a few factors.

Once, for example, the Kaibab plateau, an isolated area near the Grand Canyon, was home for a herd of deer. The herd was kept to a size equal to the plant forage on the plateau by wolves who killed and ate the deer for food. In the misguided belief that all predators were bad, men offered a bounty for each wolf killed. The wolf population rapidly disappeared. As it waned, the deer population rapidly increased. The amount of forage, however, remained unchanged. The result: many deer starved to death until the herd size again reached the limit of available food.

The balance of natural forces in communities affects life in the biosphere not only horizontally, but also vertically. Life is much more limited vertically, however, than it is laterally over the earth's surface. In general, land life is more abundant at sea level and declines in variety and numbers as one rises above this point. Many species of both plants and animals live at or near sea level on most of the continents, but the higher one climbs along the sides of mountains, the less likely are there to be many different kinds of life.

At the upper edge of the vertical development of life, mostly lichens and low shrubs, mountain sheep, and a few other hardy animals are likely to be found. Beyond this, the highest mountains are mostly sterile, except for some microorganisms.

Just as it is possible to outline life zones horizontally on earth, so can the levels of life on mountainsides be studied. Life zones on mountainsides sometimes are named for their dominant tree, but this is only one system of naming vertical divisions of life.

The vertical zones of life on mountainsides tend to overlap, and sometimes creatures and plants in one zone may be found on the edges of another, but plants, at least, are confined to their levels by sunlight, temperature, and the availability of water, and animal life often is limited by the plant life available for food. Mountain life zones, however, are not always uniform. Because mountain slopes tend to cut off rainfall by precipitating clouds as clouds are pushed upward by the shape of the mountain itself, often the side of the range away from the prevailing storms is arid. This is true of the Sierra Nevada of California. Its western slopes receive much more rain and snow than its eastern sides. The result is a much thinner distribution of life on the mountainside away from the prevailing moisture, a distribution that finally trails off into desert.

The direction of the prevailing wind, the location of mountains near the ocean, the amount of annual sunlight, and other similar factors make for great differences in the ability of many parts of the earth to sustain life. The Sahara Desert, for example, lies near the equator in Africa, but is dry, hot, and desolate. The Olympic Peninsula of Washington State, on the other hand, is in the temperate zone in the Northern Hemisphere, yet it is a rain forest. Thus it is impossible to make any general statement about biomes without first examining the specific geographic and climatic conditions to which they are subject.

One other major biome exists in the world, the ocean. It is a very special part of the earth. Not only is it vast, but much of it remains unexplored. The earth's last great frontier, it contains an immense variety of life and many different kinds of conditions. Only in the past half century have men learned much about its depths. At times the ecology of the sea seems confusing and changeable, but like the land surfaces of the world, it is as ordered and zoned as the rest of the biosphere.

Levels of life in the upper ocean, the easiest to sample, have been the first to be explored, and some knowledge of the near-surface creatures of the sea is now available. Surface life depends on how vertical ocean currents move. During winter months, storms churn the sea, transferring sediments near the bottom to the surface, where they can be consumed by creatures there. Surface life then becomes more abundant. In the spring as storms abate, food begins to sink to the bottom again, and life declines. In the fall, the churning of the ocean begins again, and a new food supply arrives. During the late fall and winter, sunlight reaching the sea is less, however, because of shorter days, and the simple plantlike creatures of the surface grow more slowly because of this.

The seasons not only affect the sea, but must be considered in studying all life zones. We tend to think the earth has four seasons, but this is true only for the temperate parts of the world. Much of the tropics have only a single yearly season, an endless summer. But this does not mean all the regions of life around the equator enjoy this state. Snow falls on the high mountains of Africa, just as it does in Europe, and may remain there for many months. In other parts of the tropics, a rainy and a dry season break the year into two parts.

The polar parts of the world usually have two seasonal changes, a long winter and a brief summer. Both spring and autumn are so short as to be almost nonexistent. In reality much of the areas around the North and South Poles are snowy deserts, where life is sparse, moisture is limited, and life may flourish suddenly and swiftly for a few weeks.

The same may be true of the deserts of the tropics, such as the Kalahari and Sahara. There the air is dry, and moisture is very limited. When rain does fall, sometimes at intervals of several years, the land suddenly blossoms. The desert, however, can also be quite cold, not as cold as the polar deserts, but it gets quite chilly at night.

These variations in climate have much to do with the migration of those creatures able to travel. As soon as the weather becomes severe, they leave for more favorable places. For some creatures, such as some migrating Arctic birds, this means nearly the entire world is their home during some part of the year. To survive, they must travel almost from the North Pole to the South Pole and back again. Other migratory creatures may make shorter journeys, but no matter how far a creature travels, it is adapting to changes in climate by migration.

Plants obviously cannot migrate, but they may make allowance for the change in weather by defensive measures, both where it is very cold and where it is very hot. When cold comes, they withdraw their sap to their roots, drop their leaves or needles and shut down as much of their aboveground life processes as possible. Desert plants store water when it becomes available, against the time when it is not present, with special stems, long roots, or root systems that cover many square feet.

Migration and other protective measures may also take place at sea, but much less is known about them than about what happens on land, simply because it is so difficult to chart the movement of undersea life with any accuracy. It is known, however, that variations in the temperature of ocean waters cause fish to move to different parts of the ocean and that whales and other sea mammals follow regular patterns of migration. Fish also move from inland streams to the ocean and return to the same streams from which they came to spawn, probably following patterns of migration set up millions of years ago.

The question of migration raises the question of the long effect of climatic changes on the evolution of life on earth, a question about which also very little is known. From studies of fossil remains of creatures long extinct, however, it is clear that many forms of life once present on the earth are not longer alive. They have disappeared, to be replaced with other forms of life. Was their passing because of changes in the general climate of the earth? Was it because of general changes in the biosphere—differences, for example, in the content of the gases in the atmosphere?

Millions of years ago, the atmosphere had less oxygen than it does today, and it seems probable that the earth also has been both warmer and colder than it is at present. How much effect these differences had on the way in which life adapted to the biosphere is less certain. Certainly, however, life has changed its form over eons of the earth's past. Dinosaurs have come and gone, the ancient fish of the sea which predated them have disappeared. In the earth's more recent history, mammals have flourished, partly at least because the climate has been more favorable to creatures able to sustain a constant interior body temperature.

The theory of evolution, as conceived by Wallace and Darwin, is based on the belief that life adapts or changes its form to be able to live with the conditions that surround it. If the conditions of the environment, the biosphere, change, then the life form must change or it will fail to survive. Change usually takes place over a long period of time, although some adaptation to climatic changes has been quite rapid. The peppered moth often is cited as an example of rapid adaptation to the environment. The moth was originally found in England with pale wings. As coal soot polluted the English cities, the peppered moth gradually changed the color of its wings to blend with its blackened surroundings, until today most peppered moths are black or a mixture of black and white. Only rarely and in rural areas of the world are pale moths found today.

Many other examples of adaptation have been found in the biosphere. Indeed, it was Darwin's discovery of small differences between birds in the Galápagos Islands of the Pacific during his voyage around the world in the *Beagle* that first led him to the theory of evolution by natural selection. Despite this, however, there remain many unresolved questions about the relation between life and the effect of the biosphere.

Do plants and animals flourish where conditions are most favorable for their growth, or are plants and animals able to adapt to changing conditions?

If life exists in fewer forms in places where the biosphere is extreme—in the Arctic and Antarctic, high on mountain slopes, deep in the sea—does this mean that the environment, rather than the ability of the creature to adapt, is most important?

Where life is most abundant, as it is around the equator, why does it take so many varied forms? Is this to provide insurance against changes in the climate? Or does life attempt to fit into every possible habitat in the biosphere?

Is it possible for a climax community to be sustained indefinitely? Or will life always be subject to changes in the environment and hence always changing in form?

This last question is one of great importance for man. Men have long assumed that they are the end product of evolution, that all that happened to life before they became the dominant species on earth was to give them control of the biosphere.

All past history in the biosphere, however, denies this assumption. No life form has ever been guaranteed permanence by the biosphere. All evolution has been a series of successions. Though man has been the most adaptable of all creatures, able to live anywhere, even in space, he can take little solace from his present state.

His hold on the world is a slippery grasp, at best. Although he has been able to control much of the environment by building shelter, by planting crops, by damming rivers, by constructing machines, in his struggle for survival he is often only a step or two ahead of disaster. Those in the technologically developed part of the world and the biosphere—principally Europe, North America, and Japan—may be able to expect a long life, but life is much shorter for those elsewhere on our planet. Disease is still prevalent. Insects and other forms of life abound and threaten man at every turn. Shelter is poor. There is little reason to expect a long life. Population is increasing more rapidly than food supplies. Man's adaptation to the biosphere in these circumstances is simply a struggle to survive.

Very probably because the world is increasingly interdependent, the adaptation of all mankind in the biosphere is dependent on how successfully he is able to cope with the mounting problems of survival. In this sense, man's problem is not only to adapt to the environment, but also to survive for the next few centuries.

Thus, the present relation between man and the biosphere is far from that of species in a climax community. Instead, it is a constant readjustment between the human race and all those things that threaten it with severe limitation or extinction.

Man lives in all the life zones of the biosphere. This increases, rather than simplifies, man's ability to survive. It makes all the world man's ecological habitat, but at the same time, it is a world which is far from secure, a world in which evolution may still be at work. Because the biosphere is still evolving, man well may not be evolution's final resolution of life within it.

—8—
POLLUTION

After World War II, residents of Los Angeles began to notice the air over their city often took on a smoky, foggy appearance, even when there was no reason for fog to form. The haze persisted through the day, sometimes growing worse in the afternoon, sometimes lasting for days. As the years passed, the smog, as it came to be called, grew worse, often filling the air with an eye-smarting mist that all but obscured the sun.

Smog in the Los Angeles basin now has become a more or less permanent part of life for its residents. Only on a few exceptional days, when the wind blows from the north across it to the sea, is the air over Los Angeles completely clear. Most of the time, especially in the late summer and early fall, when the air is still, smog may be so thick it is impossible to see the ground from a low-flying aircraft.

Los Angeles' smog happens for several reasons. The most important is the lack of city-wide mass-transit system. Almost all Los Angeles residents get from place to place by automobile. Automobiles are driven by internal-combustion engines, fueled with gasoline. Gasoline is refined from petroleum, a fossil fuel. Internal-combustion engines operate when electrical sparks are used to ignite a mixture of gasoline and air in their cylinders. The explosions that result push the pistons of their engines, which in turn drive the crankshafts. The crankshafts turn the automobiles' wheels.

All gasoline-driven engines exhaust the unburned by-products of combustion into the air. In many automobiles today, some of these are recirculated through the crankcase, but many other emissions flow through a muffler and a tail pipe into the air outside the car. The emissions are various combinations of the parts of gasoline with elements in the air. Usually they are invisible when they leave the car, but through various chemical processes all eventually contribute to smog.

Because millions of Los Angeles residents own automobiles, the amount of by-products from the operation of an automobile engine dumped into the air every day can be multiplied millions of times each day. Although the air over the city may be relatively free of pollution in the morning, by the time millions of commuters have driven to work, the amount of pollutants in the air has increased manyfold.

This would be a serious problem even under the best of circumstances. Unfortunately, the Los Angeles basin also has geographic peculiarities, which make the situation there even worse. The basin lies along a crescent of the Pacific Coast ringed by mountains 4,000 to 8,000 feet high. During the day, the prevailing wind over the city is off the ocean over the land. Warm air from the city rises until it meets this cool air. Then it is trapped and can rise no farther, for the surrounding mountains prevent it from being moved inland. But at nightfall, the air flow reverses and moves air back out to sea.

On some days when the wind is from the north over the mountains, an air inversion does not exist. On others, when the nighttime flow of air from land to sea does not take place, smog accumulates, builds up, and grows worse and worse, all but blotting out the sun. Often on such days the pollution is so severe it can cause the eyes to smart and may create difficulty in breathing.

Many solutions to Los Angeles' problems have been suggested. They range from impractical ones, such as installing giant fans on mountaintops to lift the air over them, to more realistic ones, such as banning all autos from Los Angeles. To this end, an air-pollution-control district has been established in the Los Angeles basin. It has set standards for the amounts of pollutants that cars can emit, attempted to reduce the amount of pollution from other sources, such as factory chimneys, and set up a system of air sampling and warnings to be sounded when the air grows so polluted that it becomes a danger to health.

The only real solution may have to be even more drastic: severe limitations on the use of automobiles as a means of mass transportation and the substitution of some less polluting form of getting around.

The problem of air inversions, air pollution, and its dangers of life is not confined to Los Angeles. Almost every major city

Los Angeles' predicament.

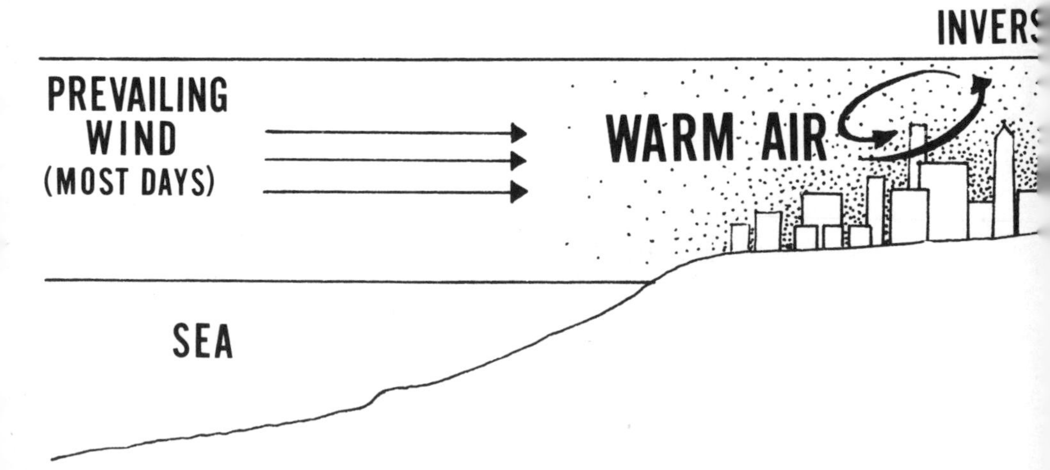

in the world now experiences smog at some time of the year, even where there is good public transportation.

In a few instances, air pollution has become so severe that it has brought human death. Three of the most famous air-pollution disasters took place in the Meuse Valley in Belgium, in Donora, Pennsylvania, and in London, England.

The Meuse Valley disaster happened in December, 1930, when a thick fog of several days' duration mixed with intense industrial air pollution from factories, much of it generated by the burning of coal. Thousands of persons were made ill, and sixty of them died.

A similar disaster took place in Donora in October, 1948, when an air inversion trapped fog and industrial emissions in a small valley. Six thousand persons became ill, and twenty persons died.

The London disaster happened in December, 1952, after a prolonged London fog fed by the smoke from thousands of coal fires, the traditional British method of heating homes. The fog

persisted for five days. An estimated four thousand people died.

In each case the dead were mostly elderly persons with serious breathing problems, but even the healthy young were affected.

Although automobile-exhaust emissions are the chief cause of Los Angeles' air pollution, this is not true of all cities. In London, the by-products of burning coal were the principal pollutants, as they were at Donora. Coal contains sulfur, which can combine with oxygen and other gases in the air to make harmful materials. Industrial pollution such as that which brought disasters to Donora and the Meuşe Valley may also contain other chemicals produced in combustion during industrial processes. Air pollutants vary from time to time and place to place, depending on their source and on the weather conditions that trap them near the earth's surface.

In all cities, the patterns of air circulation remain a problem. Because so much combustion is required to heat a city, the air above it is almost always warmer than the air around it. This warm air usually rises until it reaches higher levels of cool air. Then the warmer city air is cooled and flows away from the center of the city toward the suburbs. There it sinks toward the earth and is carried toward the center of the city again. In effect, a heat island is created over the city, which tends to recirculate its dirty air.

Air circulation in and over cities may also be affected by the shape of buildings, particularly in places like New York where there are skyscrapers. The heat island of any city must be taken into account with other local factors affecting its weather in attempting to deal with its air pollution.

The kinds of pollutants involved in this kind of a circulation pattern not only will vary with their sources, but also with the size of particles put into the air. Pollutants may be solids or liquids. If they are from one to ten microns in size, they are called particulates. (A micron is a thousandth of a millimeter, or about .00004 of an inch.)

The importance of particulates lies in their ability to be moved by the air. Particles larger than ten microns tend to fall rapidly out of the air to the ground, but particulates may move around in the atmosphere for days or weeks before this happens. The total amount of the particulates over a city may be large. On an average day, for example, particulates over the Los Angeles basin may weigh as much as 40 tons. New York's average daily load of particulates may be as much as 335 tons.

The size of particulates also has much to do with how they combine with the normal elements in the atmosphere. The smallest, those of a micron or less, may become nuclei around which water vapor condenses. They may thus contribute to the making of fog or rain.

Even larger particulates are still small enough to be breathed into the lungs. There they may lodge in the alveoli, the small air sacs through which oxygen is allowed to enter the body, to cause lung disease.

The development of nuclei around which moisture collects may have the effect of absorbing heat or of acting as catalysts for chemical reactions in the atmosphere, which might not otherwise take place. For example, the particulates left by high-flying jet aircraft may cause alterations in the delicate gas balance in the upper edges of the troposphere—at least, so some scientists believe. This has been one of the chief reasons why many American scientists oppose regular flights by supersonic aircraft at the edge of the troposphere.

Particulates come from many different sources. Automobile exhausts are only one of them. Auto exhausts may release sulfur into the air, just as does the burning of coal. Sulfur combines with oxygen in the air to make sulfur dioxide, sulfurous acid, and sulfuric acid. These various combinations of sulfur with oxygen in the atmosphere may affect plants, cause smog, reduce the amount of radiation from the sun that reaches the earth, and cause serious breathing problems in some persons. In addition, when sulfur combines with hydrogen to make hydrogen sulfide, it smells like rotten eggs. Hydrogen sulfide not only smells bad, it often coats both painted and metal surfaces with a black tarnish.

Carbon, another element involved in most combustion, combines in the air with oxygen to make carbon monoxide and carbon dioxide. Carbon monoxide is poisonous to animals, including human beings. Carbon dioxide is a normal part of the air.

The combination of carbon from combustion with hydrogen is another series of pollutants, some of which may cause cancer. Nitrogen combined with oxygen produces yet another set. Nitric oxide, nitrogen dioxide, and nitric acid are three of the most common. Nitrogen dioxide gives the air its dirty brown color when it is widespread and is often seen over California cities when smog is heavy.

The use of radioactive materials as a source of power also presents mankind with possibly serious dangers to the biosphere. Radioactive elements can damage or kill living cells. They may also alter the genetic material of the cell, making it reproduce into malignant or defective cells.

Radioactive elements have always existed on earth, but they have not been dangerous to life, because they have been widely scattered through the earth's crust. Their use to make bombs near the close of World War II was the first time man had actively mined and refined them to concentrate their radioactivity.

The purification and concentration of radioactive elements creates their great danger. Radioactive elements have half-lives —that is, the "decay," or change, at specific rates of time, from one element into another. In some cases, slight variations in the structure of the atoms of a stable, normally nonradioactive element may make it radioactive, creating an isotope. Thus, many radioactive elements also have stable, nonradioactive atoms.

The most commonly used radioactive element in the construction of nuclear bombs is an isotope of the uranium atom, uranium 238. If this isotope is separated from stable uranium, purified, and brought together in a sufficient amount, called a critical mass, it can be made to explode with great force by introducing a neutron into the mass. The neutron enters one uranium atom, causing it to come apart. As it does, it releases additional neutrons. They enter other nearby atoms. One after another at a very great speed, the uranium atoms in the critical mass fission, or break apart. It is the speed with which this process takes place that allows great amounts of energy to be freed, creating the explosive force of the nuclear bomb.

In working on this process, it became obvious to scientists that if the great energy in a nuclear bomb could be sufficiently controlled to be released a little at a time, it could provide power for the generation of large amounts of electricity. In fact, the controlled release of nuclear energy was accomplished before the first nuclear bomb was detonated, when the first nuclear reactor was built under the tennis bleachers of the University of Chicago during World War II. The first reactor was the forerunner of the boiling-water reactor, the type most in use in the world today.

A boiling-water reactor consists of a core of uranium fuel rods, which form its critical mass. The rate at which the uranium within the core is fissioned is controlled by rods of some neutron-absorbing material. As these are raised or lowered, they increase or decrease the rate at which neutrons break apart uranium atoms, releasing their energy.

The core of the reactor is surrounded by water. As the water is heated, it boils and becomes steam. The steam is fed into a turbine, which turns an electrical generator. Newer and more efficient reactors now under development will use liquid sodium instead of water to transmit heat to water, but otherwise will operate on a somewhat similar principle.

Newer "breeder" type reactors, however, will have both an additional advantage and a disadvantage. As they fission uranium, they also will "breed" plutonium. Plutonium is an element that does not occur naturally on the earth, but is created by fissioning uranium. It has a shorter half-life than uranium and hence emits more radiation, thus is more toxic to living cells. A minute amount is sufficient to cause injury or death to living cells.

The advantage of the breeder reactor is that it can replenish rapidly disappearing supplies of uranium. Its disadvantage is that it does this at a much greater risk to human life than uranium reactors. Although there are only a few plutonium reactors at work in the world today, they will increase as less and less uranium is available as a fuel source.

The advantages and disadvantages of plutonium reactors generated a great debate among scientists during the 1970's, a debate that has also involved the public. It remains to be resolved.

Plutonium and uranium reactors also produce other radioactive materials besides plutonium, some of them useful in medicine and industry, some hazardous to life. The treatment and storage of these substances, growing in number as more and more reactors are put to work, is a mounting problem in the world. Many of the substances are difficult to store. Some must be kept for hundreds, even thousands of years before they lose their radioactivity.

None of the by-products of reactors, of course, are of as great a danger as the unrestricted use of nuclear bombs in warfare. The threat of nuclear warfare continues to hang over all the world, and despite the efforts since World War II to work out some form of control of nuclear weapons, the danger has not abated.

Aboveground testing of nuclear weapons has been discontinued by all but the French and Chinese, but even so, large amounts of dangerous nuclear materials have been dumped into the biosphere. Indeed, some scientists contend that enough nuclear wastes have been released by bomb tests to bring about damage to the genes of generations as yet unborn.

Because there is no end in sight either to the testing of nuclear weapons or the construction of nuclear reactors, the atmosphere will continue for years to become a dumping ground for radioactive garbage.

It is the ocean, however, that is the earth's ultimate dump. All the waters running off the land eventually reach it, carrying with them much of the continents' pollution. In addition, man has always used the sea as the world's garbage dump. His first pollution of the waters was sewage. Cities near its shores for centuries let their human wastes run into its waters, confident that the sea would bear them far from shore. In more recent years, unwanted ships, old streetcars, useless weapons, and dangerous narcotics—a great variety of rubbish—have been dropped into the ocean, probably because men, looking out over the sea, believed it infinite and able to absorb all that was deposited into its depths.

The problem of pollution in the ocean is not only that all waterborne land wastes may contaminate the ocean, but that the sea itself has been believed capable of recycling anything it receives. This is not true. The ocean is, like the biosphere itself, a limited resource. Neither its depths nor its surface waters are capable of absorbing infinite amounts of pollutants. The currents of the sea tend to localize wastes along shorelines, but even the open sea is now contaminated, as was discovered by Thor Heyerdahl during the recent voyages of the reed boat *Ra*. Sailing from Africa to South America, Heyerdahl and his crew found visible evidence of human garbage in the Mid-Atlantic, and scientists, sampling the Atlantic farther north, have turned up minute bits of plastic, ground to microscopic size by the action of the waves. They will be present in the ocean for years, perhaps for generations.

Plastics and sewage may eventually be controlled by the careful treatment of these materials before they reach the waters of the sea, but an equally serious and as yet unsolved problem remains—that of oil spills. A large amount of the world's oil must now be removed from its underground reservoirs and carried to places far distant to be refined or consumed. The cheapest and most effective way to transport it from one continent to another is in giant oil tankers, superships so large they cannot be docked at many of the world's ports. These vessels, many of them built in Japan, now sail regularly from the Arabian Peninsula and South America to Europe and the United States, traveling well-established routes in the Atlantic.

When a supertanker flounders and sinks, as happened on March 18, 1967, immense amounts of crude oil are spread on the sea. On that day, the tanker *Torrey Canyon* was wrecked off the coast of England, spilling 36 million gallons of crude oil into the ocean. The effects of such a catastrophe are yet to be determined. The most immediate victims are shore life, particularly shorebirds. They become so coated with oil they cannot fly, and many die a slow death. But underwater life is also affected to an extent still not measured by scientists. Crude oil floats on the surface at first, but may eventually sink to the bottom, damaging or killing fish.

A disaster similar to the wreck of the *Torrey Canyon* took place in the Santa Barbara Channel a few years later, when a well being drilled into the continental shelf began to leak thousands of gallons of crude oil a day into the coastal waters of California. The oil polluted many miles of beaches before it could be shut off.

Both the drilling of offshore wells and the transportation of oil by supertankers will continue in the years ahead as petroleum becomes more and more difficult to find. The oil from the Alaska pipeline, for example, will be carried from its end to California coastal refineries by tanker, and the oil fields of the Arabian Peninsula have millions of gallons yet to be transported around Africa and through the Atlantic to Europe and North America. Despite all the precautions that are taken, there have been other oil spills since then, and undoubtedly there will be more in the years ahead.

It also seems inevitable that the seas will continue to be polluted for years to come by what happens on land. As the moving waters of the land find their way home to the ocean either as parts of moving rivers or as water vapor, they will carry with them all the waste products of industry, agriculture, and other applications of human life to the biosphere. Even if all these processes were halted tomorrow, the residue of the past two centuries of man's technological civilization would remain to be dissolved and deposited in the ocean. For years to come, the sea will remain the great dumping ground of the biosphere.

The sea is not infinite, nor, in a larger sense, is the biosphere itself. Only so far as it is able to break down the substances within it and so renew itself will the biosphere be able to survive and to nourish life. Because life itself is one of the forces of the biosphere, it, too, must cooperate in such constant recycling.

In nature, cycles provide for the reuse of waste products in many different ways. Indeed, most natural cycles are dependent on the breaking down of wastes into substances that can be used anew. Only when natural cycles are overloaded or interrupted do they pose threats to the world. As the dominant species on earth, man has had more influence on the natural cycling systems of the world than any other form of life. Through many different kinds of abuse—overpopulation, the misuse and wasteful consumption of natural resources, the failure to restore irreplaceable fuels, fibers, ores, and food—man now threatens not only his own future existence, but the future of many other species as well.

For the world, man is learning slowly and painfully, is not many places but one. The biosphere is not a group of differing environments, but a single fragile combination of elements so interdependent that one is useless without the others. The biosphere is not separated forms of life but a single entity.

And so mankind cannot continue to abuse the biosphere without counting the cost. To do so is to court disaster.

The systems of the biosphere that make life possible include the work of the sun on the waters of the sea, the weathering of the rocks and their release of oxygen, the continuation of photosynthesis, the fixation and release of nitrogen and carbon, the transport of water from sea to land and back again, and the creation and melting of ice.

All these systems are the result of the delicate balance of the biosphere. The biosphere is the product of all these systems working together. The one is dependent upon the other. Thus far in the earth's history, the systems have worked well. They have sustained and nourished life. They can continue to do so.

Yet even the biosphere is not permanent. It, too, will change. We know this most clearly because of the variation in the amount of oxygen in the atmosphere, but it is also true of other cycles. All the cycles of the biosphere are not only subject to change, but all ultimately face an end. The final lesson to be learned from the world—and, it is hoped, from this book—is that the biosphere, too, is not infinite.

We do not know how or why it will end, but several possibilities exist. Most involve the earth's relationship with the sun, for the sun, like the earth, also is finite. One day it will burn out, and its radiation—central to life on our planet—will cease. Whether this will come about through a gradual waning of heat and light, or through a series of convulsive contractions and expansions, is as yet impossible to predict.

Perhaps before either of these eventualities takes place, the earth's present balance of carbon dioxide and oxygen will have been disrupted, crippling or eliminating life. Perhaps succeeding ice ages will be so severe they will kill most life forms on our planet. Or perhaps the ice will disappear completely, filling the oceans with water, inundating most of the continents, and turning what is left of them into wastelands upon which no life will be found.

Any of these possibilities could happen. One of them is very likely to take place. Yet the death of the sun or the earth is many, many years in the future, unless, unless . . . Unless man becomes his own destroyer by rendering the biosphere uninhabitable because of nuclear warfare or by so polluting the air that he and perhaps all other life will be unable to survive.

These possibilities are nearer and far more likely than the dimming of the sun's light. They could take place within a few centuries.

Yet they need not.

Man can learn to live wisely with his world. If he learns to live wisely, he will live well. If he accomplishes both of these goals, not only he but the biosphere—the world of earth, air, fire, and water—will survive not just for centuries, but for time beyond our present comprehension.

Bibliography

Battan, Louis J., *The Nature of Violent Storms*. Garden City, N.Y.: Anchor Books–Doubleday, 1961.

Calder, Nigel, *The Weather Machine*. New York: Viking Press, 1975.

Carson, Rachel, *The Edge of the Sea*. New York: Signet Science Library, 1955.

_______ *The Sea Around Us*. New York: Signet Science Library, 1961.

Cowen, Robert, *Frontiers of the Sea*. Garden City, N.Y.: Doubleday, 1960.

Cromie, William, *Exploring the Secrets of the Sea*. Englewood Cliffs, N.J.: Prentice-Hall, 1962.

Edinger, James G., *Watching the Wind*. Garden City, N.Y.: Anchor Books–Doubleday, 1967.

Farb, Peter, *Ecology*. Life Nature Library. New York: Time-Life Books, 1963.

Flohn, Herbert, *Climate and Weather*. World University Library. New York: McGraw-Hill, 1969.

Gamow, George, *A Star Called the Sun*. New York: The Viking Press, 1964.

Marx, Wesley, *The Frail Ocean*. New York: Ballantine Books, 1967.

Pollution Primer. New York: National Lung Association, 1969.

Scientific American, *The Biosphere*. San Francisco: W. B. Freeman Co., 1970.

Van Straten, Florence W., *Weather or Not*. New York: Dodd Mead & Co., 1966.

INDEX